Developing Successful Agriculture

An Australian Case Study

FSC
www.fsc.org
MIX
Paper from
responsible sources
FSC® C013604

To all men and women of rural Australia for their dedication and contribution to advancing Australian agriculture

Developing Successful Agriculture

An Australian Case Study

Zhang-Yue Zhou

James Cook University
Townsville, Queensland
Australia

www.cabi.org

CABI is a trading name of CAB International

CABI	CABI
Nosworthy Way	38 Chauncey Street
Wallingford	Suite 1002
Oxfordshire OX10 8DE	Boston, MA 02111
UK	USA
Tel: +44 (0)1491 832111	T: +1 800 552 3083 (toll free)
Fax: +44 (0)1491 833508	T: +1 (0)617 395 4051
E-mail: info@cabi.org	E-mail: cabi-nao@cabi.org
Website: www.cabi.org	

A catalogue record for this book is available from the British Library, London, UK.

Library of Congress Cataloging-in-Publication Data

Zhou, Zhang-Yue.
Developing successful agriculture : an Australian case study / Zhang-Yue Zhou.
p. cm.
Includes bibliographical references and index.
ISBN 978-1-84593-945-8 (alk. paper)
1. Agriculture--Economic aspects--Australia--Case studies. I. Title.

HD2152.Z46 2013
338.10994--dc23

2012030952

ISBN-13: 978 1 84593 945 8

Commissioning editor: Claire Parfitt
Editorial assistant: Alexandra Lainsbury
Production editor: Lauren Povey

Typeset by AMA DataSet, Preston, UK.
Printed and bound in the UK by CPI Group (UK) Ltd, Croydon, CR0 4YY.

Contents

Foreword

The remarkable performance of Australian agriculture, amidst substantial dismantling of government supports and interventions over the last three decades, offers tremendous insight for other countries, particularly on what is required to create a successful agriculture sector when it is treated almost equally to any other business or industry in the economy. The development of such an agriculture sector is often a dream for policy makers throughout the world, and a challenge that is often given up due to lack of know-how, courage, or simply because it is considered impossible. Australia offers a great example for countries to consider in this context.

From the 1950s to 1970s, the United States provided a very good example for countries to follow on how to develop their agriculture sectors and achieve food security, and this provided the foundation for the green revolution that solved the world food problem at that time. Today, Australian agriculture can provide a great model for countries to examine as they attempt to develop successful free market agriculture sectors, even in a world where protection is still very common.

Professor Zhang-Yue Zhou is eminently suited to write this book as he grew up in rural China seeing the huge agricultural problems and government interventions there, and has then been an intense researcher and keen observer of Australian and international agriculture for several decades. Zhang-Yue also has an irrepressible zeal and enthusiasm for agriculture, the world food problem, and the well-being of the farmers around the world. It is clear from what I have read that he has put in an enormous amount of effort on this subject and has produced a monumental piece of work that is a must-read for anyone interested in the development of agriculture today, in a free market economy and world.

A major feature highlighted by the book is the institutional framework. The Australian Constitution bestows substantial powers to the states in making laws, policies and programmes for agriculture. This allows the states to tailor their institutions and policies to suit their substantially differing endowments and comparative advantage. Another very significant feature is the institution of the Productivity Commission. This is a highly independent body of experts that can conduct public inquiries on any major policy question referred to it. The Productivity Commission inquiries are open, independent, transparent and time-bound, and invite views from all constituents of society. The Commission has been used for objective evaluations on numerous major policy questions of national importance, including agricultural policies. The government is free to accept or reject the recommendations of the Commission, but the independent public inquiries become of immense value as open objective evaluations and eventually result in changes in the right direction. Some of the other important institutions highlighted are land rights, and strong agricultural services, both public and private.

Another major feature highlighted by the book is the vibrant apolitical organization of the farmers from grass-roots upwards. Australian farmers have voluntarily formed a huge number of associations, cooperatives, councils and federations for different commodities, critical resources, geographic areas, industries and other interests. These play a major role in organizing solutions and responses to various critical problems, and in empowering contact and representation upwards to the government. They are organized at the apex level into the National Farmers' Federation (NFF). Most of them are marked by dedicated and willing leadership, which is independent and apolitical, and are entrepreneurial and flexible in adapting to setting and change. They serve as great models in providing effective organization, group solutions and representation.

The book shows that Australian agriculture too was heavily regulated until the 1980s, and, as in many countries, the regulations had become complicated, cumbersome, and their effects increasingly obscure. In a series of deregulations starting in the 1980s, various price and production controls were removed, boards dismantled, and quotas and buffer-stocks done away with. The exposure to market forces had substantial implications for many, but the government largely focused on helping farmers to adjust rather than give up the deregulation process. Various rural reconstruction schemes, and specific adjustment programmes, were offered/taken up to support and allow farmers to adjust and become more resilient. It must be noted that different major umbrella supports to all people such as social security, health care and infrastructure provision also, no doubt, helped enormously to soften the impact and prevent misery.

The book highlights the enormous importance of the government effort to encourage entrepreneurship in farmers: to transform farmers from producers to business operators. The provision of information, and attracting investments to assist them, has also been very important. Efforts to encourage innovative farming by promoting innovations and breakthroughs through schemes, competitions and rewards, were also very important and gave great results. At the higher level, one of the major long-term issues is market failure in R&D resulting in declining R&D investment. A unique solution that evolved in Australia is the creation of Rural Research & Development Corporations. These are formed and supported through producer levies, industry contributions, and matching grants from the government. A special PIERD Act has been passed to provide a framework so that the corporations have independent boards, regular processes of planning and reporting, and to encourage collaboration. The structure results in promotion and support of relevant research, thereby preventing market failure and underinvestment in R&D.

The book also indicates recent efforts to foster sustainable farming, and create an environmentally conscious public and farming community. One of the major recent examples has been the effort to develop a Murray-Darling Basin Plan, which can give an adequate provision of water for the environment. The book then discusses the new challenges facing Australian agriculture including: volatile world markets; volatile weather/climate change; and the problems of farm succession, labour and skill shortage. The book indicates that getting institutions right is the most important factor, particularly to results in strong R&D, continuing education and development of public infrastructure, which are extremely important for the future. The book emphasizes the strong need to support and respect farmers, so that Australia, and the rest of the world, can have healthy agriculture sectors that can produce enough food.

Having been a keen international researcher on policies, institutions, and challenges of food and agriculture in the world for over three decades, I found this book extremely interesting and insightful, and I strongly recommend it to all readers in developing and developed countries.

Vasant P. Gandhi, PhD (Stanford)
Professor
Centre for Management in Agriculture
Indian Institute of Management
Ahmedabad, India

Preface

Agriculture remains a very important industry to developed but especially to developing countries. For those developing countries where agriculture is dominant, economic take-off still has to rely on substantial growth in their agriculture. This growth generates funds and releases labour for other industries. And improved income enables farmers to purchase more of the goods and services produced in the urban system. This then encourages urban production expansion, resulting in increased urban employment. Higher demand for workers in the urban system leads ultimately to the increased absorption of labour that becomes surplus to rural requirements.

Unfortunately, many agrarian economies in Africa, South Asia and South America are still suffering from a lack of agricultural growth. In most of these countries, institutional arrangements that could support the growth of agriculture are either weak or absent. In some societies, agriculture remains an industry that people look down on. Hence, helping poor developing countries to advance their agriculture is most desirable. Sharing the agricultural development experiences of developed countries with developing countries is one way to provide this help. This has been the major motivation for writing this book.

The other important reason for showcasing Australia's agriculture is my long-time exposure to it. I have lived in Australia for over 25 years. With my strong agricultural background and interest, I have always paid much attention to issues related to Australian agriculture. I have had extensive interactions with many people involved in Australian agriculture, from farmers, traders, consultants and researchers to policy makers in government departments and industry bodies. I admire Australian farmers, who have done wonders to develop this country's agriculture.

In a short history of a little over 220 years, Australian farmers have advanced Australian agriculture from non-existence to where it is now: one of the most advanced and efficient in today's world. This book showcases the development of

Australian agriculture and draws implications for other countries about what they could learn from the lessons and experiences of Australian agriculture. It is written chiefly for agricultural and rural policy makers in government departments and international development agencies, leaders of peak bodies of agricultural industries, researchers, agribusiness consultants and university undergraduates and postgraduates who study courses related to agricultural and rural development.

The writing of this book has benefited not only from the existing literature and publications but also from discussions and exchanges with a number of personnel who are heavily involved in Australian agriculture. The generous financial support provided by the Rural Industries Research and Development Corporation (RIRDC) in Canberra and my university, James Cook University in Townsville, made it possible for me to travel to all states and territories for extensive and in-depth discussions and exchanges with farmers, researchers, traders, industry leaders and government officials. During these visits, I also had the opportunity to inspect farm fields, trading facilities, research laboratories and so on. These personal interactions and experiences assisted me greatly in refining my understanding and thoughts about Australian agriculture.

Zhang-Yue Zhou
Townsville, Australia
1 June 2012

Acknowledgements

The writing of this book has benefited from support given to me by many individuals of various organizations. I would like to record my sincere thanks to them all.

I am most grateful to the following individuals for their generosity in spending their time to help me and in sharing their thoughts with me.

- Mr John Anderson, former Minister for Primary Industries and Energy, currently managing his farm at Mullaley, west of Gunnedah, New South Wales
- Mr Michael Badcock, vegetable farmer at Forth, near Devonport, Tasmania
- Mr Luke Bowen, Executive Director, Northern Territory Cattlemen's Association, Darwin, Northern Territory
- Mr John Brown and Mrs Pam Brown, Mr Wes Brown, 'Grenabri', Cargo, New South Wales
- Mr Craig Burns, Managing Director, RIRDC, Canberra, Australian Capital Territory
- Emeritus Professor John Chudleigh, former Principal of Orange Agricultural College and editor of the newsletter, *Analysing Agriculture*, Orange, New South Wales
- Mr Graham Coventry and Mrs Judith Coventry, 'Tobalong Tomatoes', Murray Bridge, South Australia
- Mr Rod Cox, Director, Board of Directors, Paraway Pastoral Company, Macquarie Bank, Sydney, New South Wales
- Mr Deane Crabb, Policy Manager, South Australian Farmers' Federation, Adelaide, South Australia
- Mr Ian Donges, former President of the National Farmers' Federation, now grain grower near Cowra, central New South Wales
- Mr Royce Dowling and Mr Todd Dowling, Dowling & Co., Popanyinning, Western Australia

- Mr James Doyle, Principal Regional Development Officer, North Queensland Service Centre in Townsville, Queensland Department of State Development, Infrastructure and Planning, Queensland
- Mr Peter Elliot, Senior Project Officer, Regional Services in Townsville, Queensland Department of Agriculture, Fisheries and Forestry, Queensland
- Mr Jim Farran, Manager, 'Yiddinga', Edenhope, Victoria
- Mr Wes Ford, General Manager, Primary Industries Division, Tasmanian DPIPWE, Hobart, Tasmania
- Ms Melissa Fraser, Katherine Research Station, Katherine, Northern Territory
- Mr Michael Frecker and Mrs Meredith Frecker, 'Winslow', Young, New South Wales
- Professor Vasant Gandhi, Centre for Management in Agriculture, Indian Institute of Management, Ahmedabad, Gujarat, India
- Mr David Gatenby, President of Tasmanian Farmers and Graziers Association, Tasmania
- Ms Fleur Grieve, 'Riverside Sanctuary', Northampton, Western Australia
- Dr Simon Hearn, Principal Adviser, Australian Centre for International Agricultural Research, Canberra, ACT
- Mr Tim Heffernan, 'Lomond Park', Wickepin, Western Australia
- Mr Bill Holmes, Principal Agricultural Economist, Queensland Department of Agriculture, Fisheries and Forestry, Townsville Office, Queensland
- Mrs Kerry Jonsson and Mr Greg Jonsson, Ms Kristine Jonsson, 'Jervoise Station', Greenvale, Queensland
- Mr Roger Kaus, Principal Trade and Investment Officer, Food and Agribusiness of Queensland Treasury and Trade, Townsville Office, Queensland
- Mr Tim Kelf, Manager, International Market Projects, Meat and Livestock Australia, Sydney, New South Wales
- Mr John Kerin, former Minister for Primary Industries, now retired; currently, Chair of the CRC for Poultry, Canberra, ACT
- Ms Judith Laffan, Principal Analyst, Agrifood Research Unit, Department of Foreign Affairs and Trade, Canberra, ACT
- Mr Barney Langford and Mrs Jo Langford, Manager, 'Timber Hills', Wickepin, Western Australia
- Mr Ross Leckie and Mrs Prue Leckie, 'Rocky Hill', Charleston, South Australia
- Mr Michael McBride, Presiding Member, Pastoral Board of South Australia, Adelaide, South Australia
- Mr Jock MacDonald, manager of several stations in North Queensland, now retired and residing in Townsville, Queensland
- Mr Neil Macdonald, Katherine Research Station, Katherine, Northern Territory
- Dr Bill Malcolm, Associate Professor in Farm Management, Department of Agriculture and Food Systems, the University of Melbourne, Victoria
- Dr Roger Mauldon, former Commissioner of the Industry Commission, now retired, Canberra, ACT
- Mr Bob Meaney and Mrs Julie Meaney, 'Balanda Park', Home Hill, Queensland
- Mr John Mieglich and Mrs Pam Mieglich, 'Hillside', Charleston, South Australia

- Mr Tim Mulherin, former Minister for Agriculture, Food and Regional Economies, Queensland Government, Queensland
- Dr John Mullen, previously Principal Economist at the NSW Department of Primary Industries, now retired and Adjunct Professor of Charles Sturt University, Orange, New South Wales
- Dr Roy Murray-Prior, Associate Professor in Farm Management and Agricultural Extension, School of Management, Curtin Business School, Muresk Campus, Western Australia
- Mr Gerald Neaf and Mrs Angie Neaf, Patina Winery, Orange, New South Wales
- Dr Sam Nelson, previously, Manager – Rural Affairs, NFF, now Senior Research Manager – Policy and Strategy, RIRDC, Canberra, ACT
- Dr Jammie Penm, Chief Commodity Analyst and Assistant Secretary, Agricultural Commodities and Trade Branch at ABARES, Canberra, ACT
- Mr Bruce Robinson, Robinson Rural Consulting, Perth, Western Australia
- Dr Leigh Sparrow, Soil Scientist, Tasmanian Institute of Agricultural Research (TIAR), Mt Pleasant Laboratories, Launceston, Tasmania
- Mr Jim Sullivan and Mrs Barbara Sullivan, Mr Rohan Sullivan and Mrs Sally Sullivan, Cave Creek Station, Mataranka, Northern Territory
- Professor Weiming Tian, College of Economics and Management, China Agricultural University, Beijing
- Ms Carol Vincent, Chief Executive, South Australian Farmers' Federation, Adelaide, South Australia
- Mr James Walch, 'Stewartton', Campbell Town, Tasmania
- Mr Ben Walker, General Manager, Oasis Fresh and Delroy Orchards, Perth, Western Australia
- Dr Mike Walker, agricultural consultant, Spalford, Tasmania
- Mr Wal Whiteley, 'Glenara', Molong, New South Wales
- Mr Simon Winter, Senior Research Manager, Global Challenges, RIRDC, Canberra, ACT
- Mr Don Wright, livestock and property agent, now retired and living on a farm in Millthorpe, New South Wales

Dr Roger Mauldon helped me to understand the very complex process of agricultural deregulation in Australia. He was an Associate Commissioner in 1974 and a full Commissioner in 1979 at the then Industries Assistance Commission (IAC). He continued to serve as a commissioner with the IAC and its successor, the Industry Commission (IC), until the end of 1994, when he retired. During his time there, he was involved in almost 100 inquiries, many of them relating to agricultural marketing and other agricultural and natural resource policy issues. Dr Mauldon spent hours on several occasions to explain to me the many complex issues of regulation and deregulation in Australian agriculture in the past decades. He introduced me to a number of important papers on rural policy. He also kindly gave me permission to base the section on the history of agricultural regulation on one of his papers.

Ms Sue Johnson, partnership broker (for establishing partnerships between educational institutions and business and community), Upper Taylors Arm in

New South Wales, Mr Roger Kaus, Ms Judith Laffan and Ms Stephanie Walker, Director Indigenous Relations, South Australian Chamber of Mines and Energy in Adelaide, gave me enormous help by suggesting and contacting various agricultural personnel and businesses for me to visit.

Emeritus Professor John Chudleigh, Dr Jock Fletcher, former colleague at the Orange Agricultural College, my mentor and friend, now retired and living at Mornington Peninsula in Victoria, Ms Judith Laffan and Mr Simon Winter gave me valuable encouragement to undertake this project.

Mr Peter Elliot, Mr Tim Kelf, Professor Bill Malcolm and Dr Jammie Penm have always kindly and patiently answered the numerous questions I have asked them.

Dr John Mullen kindly read and edited the manuscript. Dr Mullen's critical but constructive comments and suggestions have been very helpful in improving the manuscript. Mr Peter Elliot also read Chapter 9 and provided very valuable comments and additional information to enrich the chapter.

My university, James Cook University in Townsville, Queensland, granted me a Special Study Leave that allowed me the time to work on this project. The RIRDC provided me with financial assistance for my fieldwork.

Many of my colleagues at the James Cook University's School of Business, especially, Mrs Carmel Store and Mrs Robyn Yesberg, also gave me much assistance in my work. Discussions and exchanges with my economics colleagues have always been a great source of academic stimulation.

The love and support of my family, my wife, Jihong, my two sons, Zhou Zhou and Tian Tian, and my daughter, Dizzy, are indispensable to me and have always been greatly cherished.

Last, but not least, I would like to thank the staff at CAB International for their professional and skilful support in bringing the book to publication. In particular, I thank Ms Claire Parfitt, Commissioning Editor, Ms Alexandra Lainsbury, Editorial Assistant, and Ms Lauren Povey, Production Editor, for their guidance, patience and support during the whole process of this project. I am extremely grateful to Mrs Chris McEnnerney, who copy-edited the manuscript. Her editing skills and thoroughness are most impressive.

List of Tables

List of Figures

Abbreviations

AAA: Agriculture – Advancing Australia
ABARES: Australian Bureau of Agricultural and Resource Economics and Sciences
ABR: Australian Business Register
ABS: Australian Bureau of Statistics
ACF: Australian Conservation Foundation
ACIAR: Australian Centre for International Agricultural Research
ACT: Australian Capital Territory
AEC: Australian Environment Council
AGO: Australian Greenhouse Office
AHA: Animal Health Australia
ALP: Australian Labor Party
ANZECC: Australia New Zealand Environment and Conservation Council
AQIS: Australian Quarantine and Inspection Services
ARMCANZ: Agriculture and Resource Management Council of Australia and New Zealand
ASCC: Australian Soil Conservation Council
ASU: Amalgamated Shearers' Union
AUS$m: million Australian dollars
AWB: Australian Wheat Board
AWC: Australian Wool Corporation
AWRC: Australian Water Resources Council
BSES: Bureau of Sugar Experiment Stations
CFC: Caring for our Country
COAG: Council of Australian Governments
CONCOM: Council of Nature Conservation Ministers
CPRS: Carbon Pollution Reduction Scheme
CRC: Cooperative Research Centre

CRRDC: Council of Rural Research and Development Corporations
CSIRO: Commonwealth Scientific and Industrial Research Organization
CWFS: Central West Farming Systems
DAFF: Department of Agriculture, Fisheries and Forestry, Australian Government
DEEDI: Queensland Department of Employment, Economic Development and Innovation
DPI: Department of Primary Industries
DRIS: Debt Reconstruction with Interest Subsidy
EMDG: Export Market Development Grants
EU: European Union
FHS: Farm Household Support
FTA: free trade agreement
GAB: Great Artesian Basin
GGIP: Grower Group Innovation Project
Gl: gigalitre
GVP: gross value of production
HAL: Horticulture Australia Limited
IAC: Industries Assistance Commission
IC: Industry Commission
IOC: industry-owned corporation
IR: industrial relations
LWA: Land and Water Australia
MCFFA: Ministerial Council on Forestry, Fisheries and Aquaculture
MCU: material change of use
Ml: megalitre
MLA: Meat and Livestock Australia
MMC Co-op: Molong, Manildra and Cumnock Co-op
NFF: National Farmers' Federation
NRM: natural resource management
NRMMC: Natural Resources Management Ministerial Council
NSW: New South Wales
NT: Northern Territory
NTCA: Northern Territory Cattleman's Association
OECD: Organisation for Economic Co-operation and Development
OH&S: occupational health and safety
PBSA: Pastoral Board of South Australia
PC: Productivity Commission (Australia)
PHA: Plant Health Australia
PIERD Act: Primary Industries and Energy Research and Development Act 1989
PIMC: Primary Industries Ministerial Council
PSE: producer support estimate (formerly, producer subsidy equivalent)
QLD: Queensland
QSC: Queensland Sugar Corporation
RaL: reconfiguration of a lot
RAS: Rural Adjustment Scheme
R&D: research and development
RDC: Research and Development Corporation

RD&E: research, development and extension
RGA: Ricegrowers' Association of Australia Inc.
RIRDC: Rural Industries Research and Development Corporation
RRS: Rural Reconstruction Scheme
SA: South Australia
SAFF: South Australian Farmers' Federation
SCoPI: Standing Council on Primary Industries
SEMP: Shoreline Erosion Management Plan
SFA: Statutory Funding Agreement
SRDC: Sugar Research and Development Corporation
TAFE: technical and further education
TAS: Tasmania
TFGA: Tasmanian Farmers and Graziers Association
TPP: Trans-Pacific Partnership
VIC: Victoria
WA: Western Australia
WI: Wool International
WTO: World Trade Organization

Glossary

Agriculture: *agriculture* in this book is defined to include *farming*, forestry and fisheries.

Australian Government: the *Australian Government* is Australia's central government. It was known formerly as the Commonwealth government or the federal government. All these three names are used in this book. The Australian Government is also often referred to as the government. The Australian Government passes laws that affect the whole country. Section 51 of the Australian Constitution defines a number of issues on which the Australian Government can make laws. When 'Australian governments' are used in this book, it includes both the Australian Government and the *state* and *territory governments*.

Australian Labor Party: the *Australian Labor Party* (ALP) is Australia's oldest political party, formed in 1890. It has governed federally for about one-third of the years since federation, most recently between 1983 and 1996, under Bob Hawke and Paul Keating, and since 2007 under Kevin Rudd and Julia Gillard.

Australia's federation: Australia's formal name is the Commonwealth of Australia, which was formed in 1901 when six independent British colonies agreed to join together and become states of a new nation. The Australian Constitution enshrines the rules of government and defines how the Australian Government is to operate and on what issues it can pass laws. The birth of Australia is often referred to as 'federation' because the Constitution created a 'federal' system of government. Under a federal system, powers are divided between a central government and individual states. In Australia, power was divided between the Australian Government and the six state governments.

Broadacre: in Australia, *broadacre* refers to farms engaged mainly in growing cereals and oilseed and/or running sheep or beef cattle. According to the Australian Bureau of Agricultural and Resource Economics and Sciences (ABARES), the broadacre sector of Australian agriculture is defined to include five industry types:

- Wheat and other crops industry: farms engaged mainly in growing cereal grains, coarse grains, pulses and oilseeds.
- Mixed livestock–crops industry: farms engaged mainly in the production of sheep and/or beef cattle in conjunction with substantial activity in broadacre crops such as wheat, coarse grains, oilseeds and pulses.
- Sheep industry: farms engaged mainly in running sheep.
- Beef industry: farms engaged mainly in running beef cattle.
- Sheep–beef industry: farms engaged mainly in running both sheep and beef cattle.

Department of Agriculture: both the *Department of Agriculture* and Department of Primary Industries (DPI) have been used to name those departments that serve the agricultural industries in Australia at both the federal and state levels at different times. Over time, some other words such as forestry or fisheries have been added or deleted from these names. In this book, 'Department of Agriculture' has been used frequently, although occasionally Department of Primary Industries is also used interchangeably.

Department of Primary Industries: see *Department of Agriculture.*

Farmer: in this book, *farmer* refers to anyone who is engaged in agricultural production, including graziers or pastoralists.

Farming: *farming* is defined to include crop production and livestock production.

Grazier: see *Farmer.*

Levels of governments in Australia: there are three levels of governments in Australia. At the federal level is the *Australian Government.* At the state level are the *state* and *territory governments.* At the local level are the *local governments.*

Liberal Party: the *Liberal Party* was formed originally from a merger of the Protectionist and Free Trade parties in 1910. In coalition with *The Nationals,* it has governed federally for about two-thirds of the years since federation, most recently for nearly 12 years under John Howard (1996–2007).

Local government: *local governments* are established by the *state* and *territory governments* to take responsibility for a number of community services such as waste collection, public recreation facilities, town planning. Local governments have a legislature and an executive but no judiciary. Their powers are defined by the state or territory governments that established them. Local governments are also known as local councils. The naming conventions for local governments vary across Australia. They can be called cities, shires, towns or municipalities.

National Party: see *The Nationals.*

Paddock: in the case of Australian agriculture, a *paddock* is a block of land used for agricultural purposes. It is generally fenced for the purpose of animal production. It can also be fenced for crop production, but generally is not. The size of a paddock varies greatly depending on the purpose of land use.

Political parties in Australia: political parties are central to an understanding of how Australian politics works. They dominate state and federal parliaments, provide all governments and oppositions and frame the nature of political debate. Australia's party system is dominated by two major groups, the *Australian Labor*

Party and the coalition parties, the *Liberal Party* and *The Nationals*. In addition to these three major parties, a wide range of minor political parties exists in Australia, such as the Australian Greens and the Australian Democrats.

State and territory government: each *state government* retains the power to make its own laws over matters not controlled by the Australian Government under Section 51 of the Constitution. State governments also have their own constitutions, as well as a structure of legislature, executive and judiciary. There are six states in Australia: New South Wales (NSW), Queensland (QLD), South Australia (SA), Tasmania (TAS), Victoria (VIC) and Western Australia (WA). Territories are areas within Australia's borders that are not claimed by the states. There are ten Australian territories. Territories can be administered by the Australian Government or they can be granted a right of self-government. Self-government allows a territory to establish its own government in a similar manner to a state. Two mainland territories, the Australian Capital Territory (ACT) and the Northern Territory (NT) and one offshore territory, Norfolk Island, have been granted a limited right of self-government by the Australian Government. Outside of government, the ACT and the NT are often treated like states. The other seven territories, governed only by Commonwealth law, are: Ashmore and Cartier Islands, Australian Antarctic Territory, Christmas Island, Cocos (Keeling) Islands, Coral Sea Islands, Jervis Bay Territory and Territory of Heard Island and McDonald Islands.

The Coalition: the *Coalition* in Australian politics refers to a group of centre-right parties that has existed in the form of a coalition agreement. The Coalition partners are the Liberal Party and the National Party. The status of the Coalition varies across the Commonwealth and states. The Coalition's main rival for government is the centre-left *Australian Labor Party*.

The Nationals: originally known as the Country Party, it has held seats in the federal parliament since 1919. Its name was changed to the National Country Party in the 1970s and then to the National Party in the 1980s. In 2003, the name was changed to *The Nationals*. Federally, it has governed in coalition with the Liberal Party. National Party is used more frequently in this book.

I

Australian Agriculture: A Success Story

1 Introduction

Australian agriculture is a success story. This book provides an updated and comprehensive account of the experiences that have led to this agricultural success.

Australia's agricultural success is evidenced by the simple fact that earlier European settlers and their followers developed Australia's agriculture from non-existence 220 years ago to an agriculture that is today one of the most advanced and efficient in the world. Australian agriculture is also one of the least distorted, receiving little government assistance. Australian farmers have done a remarkable job to work the harsh environment into making Australian agriculture a world champion. Australia's successful agricultural development has benefited from the presence of strong institutional arrangements that encourage and support farmers to work the land hard and innovatively. Various other factors have also contributed to Australia's agricultural success.

Looking into Australia's agricultural development experience can be interesting, stimulating and beneficial. Comparing the agricultural development experience between Australia and one's home country can help cross-fertilization of ideas and encourage creative thinking about new solutions to old problems. Such comparison can lead to new interpretations and fresh evaluations of one's own agricultural practices with which one has long been familiar. Indeed, increasingly many scholars have realized that there is a need to look beyond the limited confines of one's own nation or region and to seek new solutions to old problems by comparing one's own practices with those used elsewhere. This book helps policy makers, researchers, agribusiness consultants and university students of other countries to relate their own agricultural development experience to that of Australia.

In this introductory chapter, in the next section we highlight the success of Australian agriculture. Section 1.2 spells out the major objectives of the book. The final section explains how the book is organized.

1.1 Success of Australian Agriculture

1.1.1 Minimal agricultural support

Australia has become one of the few countries in the world with minimal agricultural support. Agricultural support is commonly measured by the producer support estimate (PSE) (formerly producer subsidy equivalent). PSE is an indicator of the annual monetary value of gross transfers from consumers and taxpayers to agricultural producers, expressed in monetary terms as a ratio to the value of gross farm receipts. According to the Organisation for Economic Co-operation and Development (OECD), which updates and publishes PSEs annually, Australia's PSE in 2010 was only about 2%, significantly below many other countries (see Table 1.1).

Since the mid-1980s, Australia's PSE has declined continuously, attributable to the comprehensive agricultural deregulation undertaken by Australian governments. The removal of controls and regulations made various support measures redundant; hence, the reduced support payments. Today, support payments are made mainly to farmers who suffer from severe adverse weather conditions. Details about Australia's support payments can be found in the annual *Trade and Assistance Review*, produced by the Australian Productivity Commission (PC, 2011).

1.1.2 Truly free agricultural market

By the end of 2012, the agricultural market will have become completely free, thanks to the removal of all the marketing controls and regulations. Farmers are free to choose what they want to produce and how they want to sell. While this helps better resource allocation, it also improves farm operation efficiency and fosters farmers' entrepreneurial skills.

Table 1.1. Producer support estimates as a percentage of gross farm receipts, selected countries (OECD, 2012).

	1986	1990	1995	2000	2005	2010
Australia	12.96	7.95	6.45	3.31	3.66	2.23
Canada	38.13	32.93	19.66	19.96	21.31	17.71
Iceland	75.64	72.13	60.85	69.57	66.81	44.82
Japan	65.06	51.55	62.22	59.71	53.84	50.01
Korea	62.86	70.13	72.04	66.00	59.74	44.56
Mexico	3.59	15.81	–4.69	23.37	12.94	12.08
New Zealand	19.62	2.09	1.32	0.32	1.32	0.54
Norway	69.10	70.57	64.33	66.49	65.90	60.64
Switzerland	74.52	71.11	64.21	69.79	66.18	53.86
Turkey	17.42	24.56	22.55	24.15	36.88	27.90
USA	23.98	16.65	10.08	23.29	15.26	7.04
EU27	38.60	32.86	34.77	33.09	30.54	19.84
OECD – Total	37.48	31.74	31.13	32.29	27.82	18.32

1.1.3 Very competitive in the international market

Australia's agriculture is export oriented. Over 60% of its agricultural products are exported overseas. Exporting without support while other countries' agricultural exports are subsidized disadvantages Australian farmers. None the less, Australian farmers survive strong competition in the international market. Efficiency in their production and marketing operations helps them to be very competitive in the international market.

1.1.4 Quality agricultural products

Australia is known for producing 'clean and green' high-quality agricultural products, which has also helped its competitiveness in the international market, often commanding price premiums. Geographical isolation helps Australia to enjoy a very good plant and animal health status. It remains free of many major diseases and pests. This alone, however, is not sufficient to guarantee products of high standard and quality. Two other factors are also important. One is the rigorous effort by Australia to keep the country free from many exotic diseases, pests and various other threats. The other is the effort by the agricultural industries to produce high-quality products through innovative industry initiatives (e.g. traceability in beef cattle industry) to setting high enforceable industry standards and improving farming practice. Another measure that defends Australia's reputation of exports is the quality control of the Australian Quarantine and Inspection Services (AQIS).

1.1.5 Sustainable agricultural environment

Large scale land clearing by earlier settlers for agricultural use and human settlement caused substantial damage to the soil, ecosystem and biological systems. Later, increased use of chemicals and fertilizers and inappropriate use of land (such as planting and grazing on lands that were not suitable) also caused serious damage and pollution to land and watercourses. Since the middle of the 20th century, some pioneer Australian environmentalists have started to call for care of the environment. Agricultural industries also realized increasingly that they had to protect the environment on which their operations depended. Otherwise, farming could become financially unprofitable and environmentally unsustainable. Some of these problems are ongoing. Nevertheless, in the past three decades, Australian farmers have made substantial efforts and have invested significantly in rectifying the environment problem: some of which were not created by them but by their ancestors. Agricultural industries have now been rewarded for their efforts with a highly sustainable farming environment. Currently, environment protection and agricultural sustainability are a major focus of the Australian Government's agricultural policy.

1.2 Objectives

This book is written with the major objective of sharing Australia's experience in agricultural success with other countries. Specific objectives include:

- **To help developing countries to learn from Australia's experiences and lessons in its agricultural development so that they can develop their agriculture better.** Most developing economies still rely heavily on their agricultural sector. For many of them, further agricultural development and productivity improvement is a stage they must go through in order for their economy to take off. Helping such developing economies to boost their agricultural outputs further is most beneficial, not only for these countries themselves but also for the whole international community, in various ways such as reduced poverty, increased global food security, higher purchasing power for traded goods and services and an improved international peace environment. Developing countries can learn much from Australia's agricultural development and policy practices.
- **To promote to the rest of the world that Australia's agriculture is not only one of the most advanced and efficient, but also is one of the few in today's world that receives little government support.** Agriculture remains one of the most distorted industries in today's world due to heavy assistance rendered to the industry by various governments, especially those in developed economies such as the European Union (EU), the USA and Japan. Reducing or eliminating agricultural assistance has been on the agenda of various trade negotiations. But demanding that these advanced economies reduce or eliminate their agricultural subsidy has not been easy. None the less, Australian agriculture does well with little government assistance. It is relevant to demonstrate this fact to those countries that provide heavy subsidies to their agriculture. This may stimulate some advanced economies to review, and hopefully reduce, their agricultural subsidies.
- **To provide a useful and up-to-date reference about Australian agriculture.** In order to survive in a harsh natural environment for agricultural production and a challenging international trade environment for selling produce, Australian agriculture has been very responsive to external shocks. It is useful to provide updated information to readers, both inside and outside Australia, who are interested in keeping up with the dynamics of Australian agriculture.

1.3 Organization of the Book

There are ten chapters in this book. They are organized into three parts. Part I contains Chapters 1 and 2. After highlighting the success of Australian agriculture (Chapter 1), which was achieved in a unique agricultural environment (Chapter 2), Part II of the book, which is composed of Chapters 3–9, examines the key experiences, causes or factors that have helped Australia's success in agricultural development. Issues addressed in Part II include institutional arrangements for agricultural development, protection of farmers' interests, agricultural deregulation, farmer entrepreneurship, agricultural R&D, agricultural innovations and sustainable farming. The final part, Part III, consists of Chapter 10, which shows Australian initiatives in handling emerging challenges for future agricultural success.

Each of the chapters in the book, especially Chapters 3–9, can be read independently of other chapters. These self-contained chapters help the reader to gain in a single chapter a comprehensive understanding of an issue of particular interest.

References

OECD [Organisation for Economic Co-operation and Development] (2012) Producer Support Estimates (Subsidies) (http://www.oecd.org/document/0,3746,en_2649_201185_46462759_1_1_1_1,00.html, accessed 29 April 2012).

PC [Productivity Commission] (2011) *Trade and Assistance Review*, and earlier issues, Annual Report Series, Productivity Commission, Canberra.

2 Agricultural Development

This chapter serves as background reading. It will help the reader to develop a broad understanding of where Australian agriculture came from and where it is now. It will also help the reader to appreciate the achievements that Australia has made over its short history in developing its agriculture. Before we illustrate Australian agricultural development, however, it is useful to highlight the natural environment in which Australian agriculture operates and the resources on which it has based its development.

2.1 Environment and Resources[1]

Australia, officially the Commonwealth of Australia, is located in the southern hemisphere and comprises the mainland of the Australian continent, the island of Tasmania and numerous smaller islands in the Indian and Pacific Oceans. Neighbours include Indonesia, East Timor and Papua New Guinea to the north, the Solomon Islands, Vanuatu and New Caledonia to the north-east and New Zealand to the south-east.

Aboriginal settlers arrived on the mainland of the Australian continent from South-east Asia about 40,000 years before any Europeans began exploration in the 17th century. European settlement was initially through the deportation of convicts from the UK to the colony of New South Wales (NSW), founded on 26 January 1788. The population grew steadily in the following years. During the 19th century, another five, largely self-governing, Crown Colonies were established, which were Tasmania (TAS; 1825), Western Australia (WA; 1829), South Australia (SA; 1836), Victoria (VIC; 1851) and Queensland (QLD; 1859). On 1 January 1901, the six colonies became a federation – the Commonwealth of Australia was formed. Administratively, Australia is made up of six states (from the previous six colonies) and two major mainland territories: the Australian Capital Territory (ACT) and the Northern Territory (NT).

Australia is surrounded by the Indian and Pacific Oceans and is separated from Asia by the Arafura and Timor Seas. It has a land mass of almost 7.7 million km^2. It is the sixth largest country by total area in the world. While Australia is the world's smallest continent, it is often considered the world's largest island.

2.1.1 Climate

Australia has a relatively harsh and variable climate. It has a wide range of climatic zones, from the tropical regions in the north, the arid expanses of the interior and the temperate regions in the south, to the subtropical maritime climate along the coast. The temperature can range from above 50°C to well below zero. Extremely cold temperatures as recorded in other continents do not occur in Australia because of the lack of mountain ranges and the expansive surrounding oceans.

Australia is widely known as a very dry continent. Eighty percent of the country has a median rainfall of less than 600 mm/year and 50% less than 300 mm. The lowest rainfall median, in the Lake Eyre region of SA, is only about 100 mm. Another low rainfall area is the Gile–Warburton region in WA, which has a median rainfall of about 150 mm/year.

A vast region of Australia, extending from the west coast near Shark Bay, across the interior of WA and SA to south-west QLD and north-west NSW, has an annual median rainfall of less than 200 mm. Regions with rainfall medians above 600 mm/year include the east coast of QLD, western TAS and the Snowy Mountains area in NSW. Not only does Australia have extreme variability in rainfall across the continent but also it has extreme variability in rainfall from year to year.

December to February is Australia's summer, while June to August is its winter. In between is spring, from September to November, and autumn, from March to May. In latitudes less than 20°S, the reference is not so much to summer and winter but to the wet season and the dry season.

Average annual temperatures in Australia range from 28°C along the Kimberley Coast in the extreme north of WA down to 4°C in the alpine areas of south-eastern Australia.

Most of the Australian continent receives more than 3000 h of sunshine a year, or nearly 70% of the possible total. Central Australia and the mid-west coast of WA receive sunshine slightly in excess of 3500 h, while at the other end of the scale totals of less than 1750 h occur on the west coast and highlands area of TAS. In southern Australia, the duration of sunshine is greatest around the month of December, when the sun is at its highest elevation, and lowest in June, when the sun is low. In tropical northern Australia, sunshine is generally greatest about August–October, prior to the wet season, and weakest about January–March, during the wet season.

Australia's climate is influenced significantly by ocean currents, including the Indian Ocean Dipole, the El Niño-Southern Oscillation and the seasonal tropical low pressure system that produces cyclones in northern Australia. Australia frequently experiences many weather-related extreme conditions, such as heatwaves, severe droughts and floods, bushfires, dust storms, severe hail, thunderstorms and gales.

2.1.2 Topography

Australia is the lowest, flattest and, apart from Antarctica, the driest of the continents. The age of landforms in Australia is generally measured in many millions of years. This gives Australia a very distinctive physical geography. Most of the continent is at a relatively low elevation, with less than 1% of the country above 1000 m elevation. Elevations exceeding 2000 m are found only in the Snowy Mountains of NSW, with the highest peak being Mt Kosciuszko (2228 m).

The mainland continent is generally divided into three large areas, although there are no defined boundaries between them:

- the Western Plateau;
- the Central Lowlands; and
- the Eastern Highlands.

Much of the Western Plateau is relatively flat. There are, however, numerous more rugged areas near the coastal boundaries of the plateau, including the Kimberley region and the Hamersley Ranges in WA, as well a number of relatively isolated ranges in central Australia (such as the MacDonnell and Musgrave Ranges) and individual mountains, of which Uluru (Ayers Rock) is probably the best known.

The Central Lowlands stretch from the Gulf of Carpentaria through the Great Artesian Basin to the Murray–Darling Plains. Most of this area is flat and low lying. The main exception occurs in SA, where relatively recent faulting has occurred and the area takes the form of a number of blocks that have been moved up to form a series of ranges (e.g. the Flinders Ranges and the Adelaide Hills), with the down-faulted blocks in between forming plains, some of them submerged (e.g. Spencer Gulf). Much of the Central Lowlands is occupied by the Great Artesian Basin, which consists of sedimentary rocks that hold water which enters in the wetter Eastern Highlands.

The Eastern Highlands, stretching along most of the length of the east coast, are characterized over much of their length by a steep escarpment on the coastal side, a series of high plateaus and then more gentle sloping towards the inland. While the highest elevations (over 1800 m) are found in the Snowy Mountains and Victorian Alps, many of the plateaus further north in NSW exceed 1000 m elevation. In QLD, however, 1000 m is reached in only a few locations, and the highlands are generally less prominent.

As a result of the plateau-like nature of much of the Eastern Highlands, the Great Dividing Range, which separates rivers flowing to central Australia or the Murray–Darling Basin from those flowing to the Pacific Ocean or Bass Strait, is not very pronounced in most locations. Because of the close proximity of the

Great Dividing Range to the eastern seaboard, the coastal strip is fairly narrow and is rarely more than 100 km wide. The strip is flat in some places, but quite hilly in many others.

Australia's topography, together with the harsh climate conditions and a large area of desert, dictates that a large portion of this continent is not suitable for agricultural purposes. Most of Australia's agricultural operations are concentrated in the south-east and south-west corners. There are also agricultural operations along the seashore, especially the eastern coastline.

2.1.3 Rivers and lakes

The rivers of Australia may be divided into two major classes: those of the coastal margins, with moderate rates of fall, and those of the central plains, with very slight fall. Australia's longest river system, the Murray–Darling, drains part of QLD, most of NSW and northern VIC and a section of SA, finally flowing into the arm of the sea known as Lake Alexandrina, on the SA coast. The length of the Murray is about 2520 km, while the longest branch of the combined Murray–Darling system, with its headwaters in the Culgoa catchment, is about 3370 km.

Most of the east coastal rivers are short, the exceptions being those rivers which penetrate the coastal escarpment, such as the Burdekin and Fitzroy in QLD and the Hunter in NSW. The north-east of WA also has a number of short coastal rivers.

In addition to those rivers which form part of the Murray–Darling Basin, western QLD has a number of inland-flowing rivers, such as the Paroo, Bulloo, Diamantina and Cooper Creek. These rivers do not reach the sea but drain into Lake Eyre or dissipate without reaching any other river system.

A number of river systems reach the tropical or subtropical coast. Many of these are of considerable length, such as the Mitchell, Gregory and Leichhardt in northern QLD, the Daly and Victoria in the NT and the Ord, Fitzroy, Ashburton, Fortescue and Gascoyne in WA. All of these rivers have extremely large variations in flow between wet and dry seasons, arising from the great seasonal rainfall variations typical of this region, and some only flow intermittently. The Mitchell, whose annual discharge of about 12 km^3 rivals the Murray–Darling as Australia's largest river system in terms of volume, has discharges in February and March about 100 times those of July.

There are many lake types in Australia. The largest are salt lakes, which are, or were, drainage sumps from internal rivers. For most of the time, these lakes are beds of salt and dry mud. Lake Eyre, which was filled three times in the whole of the 20th century and recently in 2011, is the largest of these (9500 km^2). Other large salt lakes include Lake Torrens (5745 km^2) and Lake Gairdner (4351 km^2). There are also some other natural lake types. These include coastal lakes formed by the damming of valleys by marine sediments, fault angle lakes (such as Lake George near Canberra), volcanic lakes (mostly in VIC, south-eastern SA and QLD) and glacial lakes (most common in TAS but also found in the Snowy Mountains). Many of these lakes are permanent but some, such as Lake George, dry out during drought periods, and all are small compared with the inland salt lakes. Australia has no natural, unmodified, permanent freshwater lake larger than 100 km^2.

Many artificial lakes, or lakes expanded by artificial means, also exist in all states and territories. The combined Lakes Gordon and Pedder in south-western TAS are the largest of these, both in surface area (513 km^2) and volume (11,320 megalitres (Ml)), while other very large artificial lakes include Lake Argyle on the Ord in northern WA (5720 Ml) and Lake Eucumbene in the Snowy Mountains Scheme (4870 Ml).

2.1.4 Water resources

In Australia, the major source of water is rainfall. Australia also has one of the world's large aquifer systems, the Great Artesian Basin (GAB), estimated at 1.7 million km^2 and with a storage volume of 8,700,000 gigalitres (Gl). Each year the GAB supplies 570 Gl of water for a variety of uses dominated by pastoral enterprises.

2.1.5 Land and soil

Before European settlement, Australia's land was covered by native vegetation. Australia's soils are old and shallow and are susceptible to degradation by disturbance to native vegetation. Since European settlement, however, vast areas of native vegetation have been cleared for human settlement and the expansion of agriculture. Clearance of native vegetation has caused some serious environmental problems such as salinity, which in turn has resulted in lost agricultural production and poor water quality for communities relying on rivers for water. Loss of native vegetation has also been a threat to biodiversity (Mullen, 2001).

2.1.6 Air

Air pollutant levels are not considered to be high in urban Australia relative to other world cities.

2.2 Origin and Evolution of Australian Agriculture

Having highlighted Australia's natural environment and resources, we now review how Australian agriculture started, how it evolved in the harsh environment and where it is now.

Australian Aboriginals are hunters and collectors. When the Europeans came to Australia to settle in 1788, there was no indigenous agriculture. Early European settlers and their followers did a most remarkable job to establish Australian agriculture from non-existence to a modern agriculture today. Looking back on the development of Australian agriculture since 1788 is itself an interesting and exciting exercise. However, only a brief account is presented here, which is based partly on

Shaw (1982, 1990). Those who are interested in more detailed accounts of Australian agricultural history may find the following useful: Davidson (1981), Shaw (1982, 1990), Ashton (1988), Gruen (1990) and Henzell (2007).

In 1787, George III, King of England, gave instructions to Governor Phillip to establish a new colony in Australia. The first settlers, consisting of convicts and free men, came ashore at Sydney on 26 January 1788. They faced the challenge of producing an adequate amount of food for themselves in the new colony. They tried to grow crops and raise animals but the job was to prove difficult, and at times disastrous.

The first attempt to grow wheat was on 3.6 ha of land at what are now Sydney's Botanic Gardens. But the experiment was a failure. Seed brought on the voyage had been partly ruined by weevils. Much of the viable seed was destroyed by mice and ants after sowing. An attempt to grow vines at about the same time also ended in disappointment, due to plant diseases.

There were also other obstacles. The new environment was very different to the one the convicts and their keepers had known in England. Soil and husbandry were both poor. Farming tools were scarce and of poor quality, but their supply was expensive. These problems were made worse by the fact that few of the colonists had any farming experience (Clark, 1992, pp. 22–24).

May 1790 saw the colony on the verge of starvation. But several supply ships arrived and, by the end of that year, a small but promising quantity of wheat and barley was harvested. From this time on, colonial farming expanded slowly. Grants of land were made to ex-convicts, the first of which went to James Ruse at Parramatta in 1791. Ex-servicemen were next to receive land, and finally, in 1793, the colony's civil and military officers were granted land.

Farming remained a difficult and hazardous occupation for many years. Small farmers in particular suffered many hardships. Their situation was made worse by the unfair dealings of the colony's officers and, later, various merchants who monopolized the market for all sorts of items. They sold goods to small settlers for high prices and, whenever possible, paid them low prices for their produce.

Despite the hardships, the number of farmers in the colony continued to grow as increasing numbers of freed convicts, ex-marines and a few free immigrants were given land. Settlement therefore spread and larger quantities of wheat, maize and other crops were produced.

From the 1820s, sheep grazing, or pastoralism, became more important than cropping. Colonial governors and officials back in England began to promote this industry, hoping that the colony would be able to pay for itself and perhaps become profitable.

Permanent settlement in VIC began in the 1820s and was based largely on pastoralism. So, too, was QLD's settlement from 1840. Both of these colonies were part of NSW until the mid-19th century.

After SA was founded in 1836, some of its colonists also took up pastoral activities and agriculture was to become the most important activity in that colony. In WA, which was first settled in 1826, development in both cropping and grazing was slow.

As settlement expanded, increasing areas of forest and scrub were cleared away. But in eastern Australia, much of this land was taken up illegally by pastoralists. This began on a large scale during the 1830s in NSW, and the people involved became known as 'squatters'. They 'locked up' the best land.

Other factors also contributed to slowing down the expansion of farming. Transport was inadequate, so farmers were forced to stay near markets or in areas where their produce could be shipped along rivers. There were also few mechanical inventions until the middle of the 19th century. For some time, therefore, farming remained backward and was carried out largely by poorer settlers.

In 1851, gold was discovered in Australia. Many farmers left the land to go prospecting. By the late 1850s, when gold became more difficult to find, some farmers gradually returned to the land. By this time, Australia's population had increased greatly due to the flood of migrants who came in search of gold. In 1840, the country's population was about 190,000; in 1850, it reached 405,000 and by 1860, it exceeded one million. As a result of the increasing population, farming was stimulated.

Many immigrants and other colonists began demanding that colonial governments let small-scale farming begin on government land. But most of the best land had been 'locked up' by squatters. Efforts were made by colonial governments to redistribute lands to many new small farmers. However, a great deal of the land distributed to these newcomers was not suitable for agriculture: transport facilities were inadequate and expensive for farm produce; most settlers did not have enough capital to develop their land properly; and few of them were skilled in farming or fully aware of the peculiarities of Australian agriculture. Farming life was hard and often unrewarding.

The depression of the late 1880s and 1890s ruined many farmers. Some small farmers left the land and drifted into towns and cities in search of work. Squatters, too, faced hard times. Having borrowed money to buy their properties, many were unable to pay off their debts. Their lands were sold to even wealthier pastoralists or ended up in the hands of banks and financiers.

By the turn of the 20th century, the number of farmers had increased, as had the amount of produce grown. Increases in production were largely the result of developments in science and technology. During the 19th century, particularly in the latter half of the century, various agricultural machines were invented in Australia. These machines allowed farmers to expand further inland. From the 1860s, a number of agricultural colleges began to undertake scientific research and to train people in the more 'modern' methods of farming. Agricultural researchers also began to breed new varieties of wheat that could resist drought and plant diseases. Major experiments with irrigation also began in this period. All these agricultural innovations had an impact on farming in the last quarter of the 19th century. But it was not until the 20th century that they had their greatest impact.

After the depression of the 1890s, governments looked to farming as a way to boost Australia's economy. During the first decade of the 20th century, various pieces of legislation were passed in most states to promote small-scale farming. Many people took the opportunity to move to the land.

The ways in which farming was conducted were gradually changing. More and more farmers were using horse-drawn machines. Departments of agriculture, most of which were formed late in the 19th century, and agricultural colleges were slowly beginning to influence farming practices. New varieties of wheat also increased the size of harvests, while allowing farmers to move into less fertile areas. New methods and equipment had also expanded the range of crops that farmers could grow.

Still, farming continued to be difficult and, at times, unprofitable. Even in good seasons, women and children had to work long hours without being paid. At harvest times, country schools were often empty as children were used as harvest labour.

The First World War (1914–1918) interrupted the general developments in farming. However, wartime needs for food, particularly by the British, stimulated some farming sectors. For example, during the war, butter was exported overseas from NSW for the first time in that state's history.

As a result of the First World War, the federal and state governments became involved more directly in controlling and regulating the way farming was carried out. They also began to create new systems for transporting and marketing products. From the 1920s, wheat began to be handled in bulk rather than in bags.

Recession in the 1920s turned into the depression of the 1930s. The Great Depression affected some farming activities more than others. In general, wheat farmers and orchardists suffered the most, and many farmers were forced off the land in the 1930s. By the second half of that decade, various government boards were set up to 'reconstruct' some farming industries.

During the Second World War (1939–1945), shortages of labour and materials disrupted farming. The Australian Government took various measures in order to maintain food supplies for troops fighting overseas and for civilians at home. Certain activities, such as vegetable growing, were given priority.

After the Second World War, Australian farming entered a long period of change and growth: the average size of farms became larger; farms became increasingly mechanized; and farm machinery became more powerful. At the same time, the number of farms began to decrease. Small farms that were unable to 'modernize' could not compete. Those failed small farms were incorporated into more efficient farms. The adoption of more expensive equipment and procedures also reduced the need for labour. Thus, the number of people employed on farms also began to shrink.

Until the 1970s, Australian agriculture remained heavily controlled and regulated (Chapter 5 provides details of the ways used to control and regulate). Agricultural deregulation began in the early 1970s and slowly gathered momentum. In the last two decades of the 20th century, farming became further deregulated. Farmers became more and more exposed to competitive markets. Adjustment assistance was provided to those who experienced hardship as a result of deregulation. In some cases, some less viable farmers were encouraged and assisted to exit farming. This led to a further drop in the number of farms but larger and more efficient farming enterprises. Today, Australia's farmers are among the most efficient and competitive in the world.

2.3 Australian Agriculture Today

This section provides an overview of Australian agriculture at the beginning of the 21st century. It highlights the use of agricultural land, the number of agricultural establishments and the number of workers employed; farm ownership; farm inputs and costs; agricultural outputs; agricultural terms of trade; and the contribution of agriculture to the national economy.

2.3.1 Agricultural land

According to the Australian Bureau of Statistics (ABS, 2008b, pp. 6–7), approximately 425,449,000 ha or 55.3% of Australia was occupied by agricultural businesses in 2006–2007. The remainder consisted of land occupied by government agencies, reserved lands and unoccupied lands. Of land occupied by agricultural businesses, 67.9% was used for grazing on land other than improved pasture, 16.2% for grazing on improved pasture, 8.9% for crops, 3.4% for conservation and 3.2% for other uses (including forestry).

At the state level, the lowest proportion of land occupied by agricultural businesses was in TAS (24.3% of the state area), while the highest was in QLD (83.1% of the state area). The percentage of land managed by agricultural businesses that was used for grazing on improved pasture ranged from 7.3% in WA and 7.4% in the NT to 43.8% and 52.7% in VIC and TAS, respectively. Areas used for grazing on land other than improved pasture ranged from 9.8% of agricultural land in VIC to 83.3% of agricultural land in the NT. The percentage of agricultural land used for crops ranged from less than 1% in the NT to 38.6% in VIC. In TAS, 8.4% of land managed by agricultural businesses was set aside for conservation, well above the national average of 3.4%. Similarly, the percentage of land managed by agricultural businesses that was used for other purposes, including forestry, was also higher in TAS: 10.0% compared to 3.2% nationally (ABS, 2008b, p. 7).

2.3.2 Number of farms and rural workers

In 2009–2010, there were 134,184 agricultural establishments in Australia, employing 369,200 people. The total number of agricultural establishments has declined from 197,050 in 1967–1968 to the present 134,184: a decline of 32%. The number of workers employed has also declined from 458,000 to 369,200: a decline of 19% during the same period (Table 2.1). In the mid-1960s, the contribution of rural employment to the national total was about 10%. This contribution dropped to a mere 3% by the beginning of 2010.

Contributing family workers are declining, in both absolute and relative terms. This may be due to the fact that family members may have been working off-farm or some children may have left home for study or work elsewhere. It is interesting to note that the number of owner-workers is declining, while that of wage earners is increasing. The decline in the former is related to the decline in the

Table 2.1. Number of agricultural establishments and rural employment in Australia (based on ABARES, 2011, p. 23).

	Agricultural establishment	Rural employment							National total employment
		Employers and self-employed		Wage and salary earners		Contributing family workers		Rural total	
	Number	Thousands	Percent of total	Thousands	Percent of total	Thousands	Percent of total	Thousands	Thousands
1967–1968	197,050	N/A	N/A	N/A	N/A	N/A	N/A	458.0	4,989
1969–1970	192,550	N/A	N/A	N/A	N/A	N/A	N/A	452.0	5,285
1979–1980	179,084	N/A	N/A	N/A	N/A	N/A	N/A	404.0	6,187
1989–1990	163,416	N/A	N/A	N/A	N/A	N/A	N/A	425.5	7,822
1999–2000	146,371	220.5	50.6	192.8	44.2	23.3	5.3	436.0	8,835
2000–2001	140,516	211.0	48.9	198.5	46.0	22.0	5.1	431.6	9,018
2001–2002	135,377	211.8	48.4	213.0	48.7	13.0	3.0	437.6	9,140
2002–2003	132,983	178.8	48.3	181.3	48.9	10.5	2.8	370.4	9,380
2003–2004	130,526	172.0	46.9	185.3	50.5	9.5	2.6	366.8	9,526
2004–2005	129,934	166.5	46.6	181.3	50.8	9.8	2.7	357.1	9,786
2005–2006	154,472	161.8	46.5	176.3	50.6	10.0	2.9	348.1	10,088
2006–2007	150,403	166.8	47.6	176.5	50.4	7.0	2.0	350.4	10,374
2007–2008	140,704	171.0	48.3	173.8	49.1	7.3	2.0	354.0	10,684
2008–2009	135,996	176.5	48.7	173.5	47.9	7.8	2.1	362.4	10,892
2009–2010	134,184	173.0	46.9	188.5	51.1	7.5	2.0	369.2	11,027
2010–2011	N/A	N/A	N/A	N/A	N/A	N/A	N/A	351.4	11,355

number of agricultural establishments. In the future, both the number and the share of wage earners in Australian agriculture are expected to grow.

2.3.3 Industry distribution of agricultural establishments

Among those 150,403 agricultural establishments in 2006–2007 (the latest year with data of greater details on industry distribution), about 10% (15,685) are agricultural businesses where the main industry is one other than agriculture. The remaining 90% (134,718) are agricultural businesses with agriculture as their main industry.

Sheep, beef cattle and grain farming account for an overwhelming majority of all agricultural establishments (91,467, 61%). Around 9% (13,580) are engaged in fruit and tree nut growing. The number of dairy cattle farms is 6973 and accounts for almost 6% of the total agricultural establishments. Mushroom and vegetable growing account for about 3.5% (5065). Further details about the distribution of all the agricultural businesses in various industries are given in Table 2.2.

2.3.4 Farm ownership

An important feature of Australian agriculture is the high incidence of owner-operators. Evidence from ABARES's farm surveys of a number of different agricultural industries over recent years indicates that more than 95% of Australian agricultural establishments are family owned and operated (owned and operated by related individuals). For example, for the broadacre (covering farms engaged mainly in growing cereals and oilseed and/or running sheep or beef cattle) and dairy industries, which together account for 68% of all commercial-scale farm businesses in Australia, ABARES's most recent survey results estimated 98% of farm businesses in these industries were family owned and operated in 2010–2011 (P. Martin, ABARES, 26 April 2012, personal communication). There is corporate farming in Australia but its share-out of total agricultural establishments is relatively small.

According to the ABS (2011), the total number of agricultural businesses in Australia on 31 December 2010 was 135,600.[2] Of this total, 133,600 agricultural businesses, or 99% of Australian agricultural businesses, were entirely Australian owned, 1300 agricultural businesses had some level of foreign ownership and 732 had an unknown ownership status. Of the businesses reporting some level of foreign ownership, 139 businesses had less than 10% foreign ownership, 330 had between 10% and 50% foreign ownership and 824 had greater than 50% foreign ownership.

In absolute terms, the states that have larger numbers of agricultural businesses reporting foreign ownership being more than 50% are NSW/ACT (313), followed by WA (117) and VIC (107). Proportionally, the NT had the highest level of foreign ownership of agricultural businesses, with 3.2% of its businesses (20) reporting some level of foreign ownership. QLD had the lowest level of foreign ownership of agricultural businesses, with 0.5% of its businesses (132) reporting some level of foreign ownership.

Table 2.2. Industry distributions of agricultural establishments, 2006–2007 (based on ABS, 2008b, p. 30).

Industry	Number	Percent of total
011 Nursery and floriculture production	***2,239***	***1.49***
0111 Nursery production (under cover)	421[a]	
0112 Nursery production (outdoors)	780[a]	
0113 Turf growing	301[a]	
0114 Floriculture production (under cover)	258[a]	
0115 Floriculture production (outdoors)	479[a]	
012 Mushroom and vegetable growing	***5,065***	***3.37***
0121 Mushroom growing	97[b]	
0122 Vegetable growing (under cover)	913	
0123 Vegetable growing (outdoors)	4,055	
013 Fruit and tree nut growing	***13,580***	***9.03***
0131 Grape growing	6,039	
0132 Kiwifruit growing	46[b]	
0133 Berry fruit growing	465[a]	
0134 Apple and pear growing	780	
0135 Stone fruit growing	1,193	
0136 Citrus fruit growing	1,388	
0137 Olive growing	398[a]	
0139 Other fruit and tree nut growing	3,271	
014 Sheep, beef cattle and grain farming	***91,467***	***60.81***
0141 Sheep farming (specialized)	12,150	
0142 Beef cattle farming (specialized)	44,957	
0143 Beef cattle feedlots (specialized)	243[b]	
0144 Sheep–beef cattle farming	8,501	
0145 Grain–sheep or grain–beef cattle farming	14,131	
0146 Rice growing	127[a]	
0149 Other grain growing	11,358	
015 Other crop growing	***6,973***	***4.64***
0151 Sugar cane growing	3,975	
0152 Cotton growing	526[a]	
0159 Other crop growing n.e.c.[c]	2,472	
016 Dairy cattle farming	***8,921***	***5.93***
0160 Dairy cattle farming	8,921	
017 Poultry farming	***1,296***	***0.86***
0171 Poultry farming (meat)	797	
0172 Poultry farming (eggs)	499	
018 Deer farming	***184***	***0.12***
0180 Deer farming	184[b]	
019 Other livestock farming	***4,992***	***3.32***
0191 Horse farming	3,019	
0192 Pig farming	894	
0193 Beekeeping	30[b]	
0199 Other livestock farming n.e.c.	1,049[a]	
All agriculture	***134,718***	***89.57***
All other industries[d]	***15,685***	***10.43***
Total all industries	150,403	

Notes: [a]Estimate has a relative standard error of 10% to less than 25% and should be used with caution; [b]estimate has a relative standard error of 25–50% and should be used with caution; [c]not elsewhere classified; [d]consists of agricultural businesses where the main industry is one other than agriculture.

2.3.5 Farm inputs and costs

Total farm costs in 2010–2011 were AUS$36,576m (ABARES, 2011, p. 75). Farm costs consist of a number of components (Fig. 2.1). Among them, interest paid and depreciation are the two major ones, each being about 13.5% in 2010–2011. The next major components are seed and fodder, marketing, repairs and maintenance, wages and other materials and services, each being slightly over 10%. Fuel, fertilizer and chemicals account for around 4–6%. Over time, the share-out of total costs for each of the major components is relatively stable. The one with the largest variation is interest, ranging from about 8–14% in the past two decades. The change in the cost share of some material items such as fuel, chemical and fertilizer use seems to be related closely to global energy prices.

Worth particularly noting is the use of fertilizers and chemicals in Australian agriculture. The use of all three major fertilizer elements (phosphate, nitrogen and potash) has declined in recent years in total quantity terms (Table 2.3). Per hectare use of each of the three elements has also declined. The reasons responsible for this decline may include the following: (i) public demand for reduced fertilizer use to cut down runoffs to the water system, although the impact is likely to be small; (ii) more precise application of fertilizers, thus reducing wastes; and (iii) economic and seasonal conditions affecting the use of fertilizers. In poor seasons and times of low prices, farmers would use less fertilizer. There were severe adverse weather conditions in the first decade of the 21st century. None the less, the long-term trends point to a decline in per hectare use of fertilizers. It would be interesting to see the application level of fertilizers in years after 2009 when data become available.

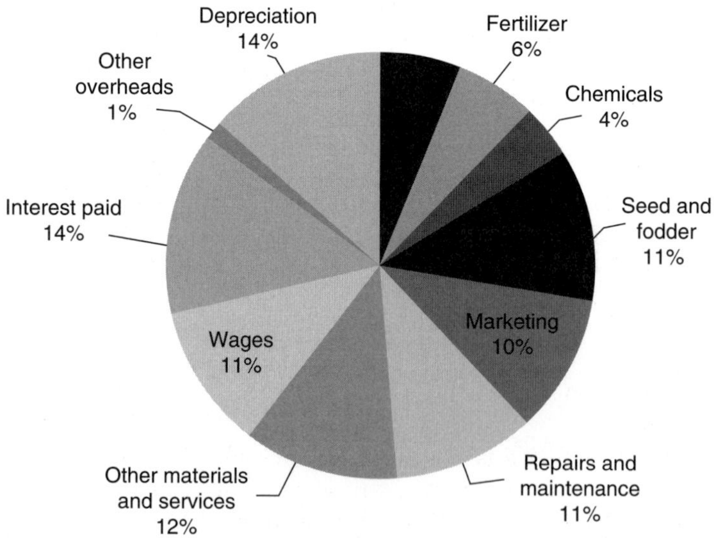

Fig. 2.1. Components of Australian farm costs 2010–2011 (based on ABARES, 2011, p. 75).

Table 2.3. Fertilizer use in Australian agriculture (1982–2009) (based on ABARES, 2011, p. 84).

	Phosphate (P_2O_5)		Nitrogen (N)		Potash (K_2O)		Cropping area
	Total (kt)	kg/ha	Total (kt)	kg/ha	Total (kt)	kg/ha	(m ha)
1982	762	44	250	15	137	8	17
1985	685	38	344	19	141	8	18
1990	579	37	439	28	145	9	16
1991	680	46	462	31	142	10	15
1992	782	47	488	29	147	9	17
1993	770	47	565	34	176	11	17
1994	923	59	583	37	220	14	16
1995	965	54	671	37	231	13	18
1996	985	49	825	41	206	10	20
1997	1090	54	839	42	255	13	20
1998	1039	48	979	45	232	11	22
1999	1055	48	1082	49	216	10	22
2000	1107	49	951	42	217	10	23
2001	1187	54	1034	47	242	11	22
2002	1078	50	980	46	215	10	21
2003	1019	43	933	39	239	10	24
2004	1107	45	1056	43	259	11	24
2005	1041	46	952	42	222	10	23
2006	978	45	858	39	222	10	22
2007	982	41	849	36	227	10	24
2008	818	33	835	34	215	9	25
2009	641	26	850	35	158	6	24

On the other hand, chemical use has been increasing (Table 2.4). In the 1960s, chemicals were used sparsely. Since the 1960s, the wider and increased application of herbicides, fungicides and plant growth regulants, insecticides to control and prevent diseases, pests and weeds in crop production has lifted crop yields dramatically (W. Brown, Cargo, NSW, 20 November 2010; T. Heffernan, Wickepin, WA, 12 February 2012, personal communication). This is partly in response to an increase in reduced tillage farming. Similarly, the use of animal health products has improved livestock production significantly (R. Sullivan, Mataranka, NT, 4 December 2010, personal communication). The use of chemicals is likely to increase further.

2.3.6 Outputs of cropping and livestock industries

In 2010–2011, with total costs of AUS$36,576m, the gross value of farm production generated was AUS$48,674m, giving a net value of farm production of AUS$12,098m. Crops contributed to some 56% of the gross value of farm production (grains and oilseeds: 26%; other crops: 30%) (Fig. 2.2). Livestock contributed the remaining 44% (livestock slaughtering: 29%; livestock products: 15%).

Table 2.4. Australian sales of agricultural chemicals, by product type (ABARES, 2011, p. 92).

	Herbicides	Fungicides, plant growth regulants	Insecticides	Animal health products
	AUS$ (thousands)	AUS$ (thousands)	AUS$ (thousands)	AUS$ (thousands)
1975	25,594	8,237	13,424	N/A
1980	94,301	17,551	34,986	88,552
1985	207,226	30,165	52,659	155,542
1990	398,336	57,856	115,108	267,017
1995	503,700	86,400	189,300	271,500
2000	883,677	155,525	326,686	511,580
2005	908,519	185,120	321,139	620,744
2010	1,192,623	182,085	329,894	820,437

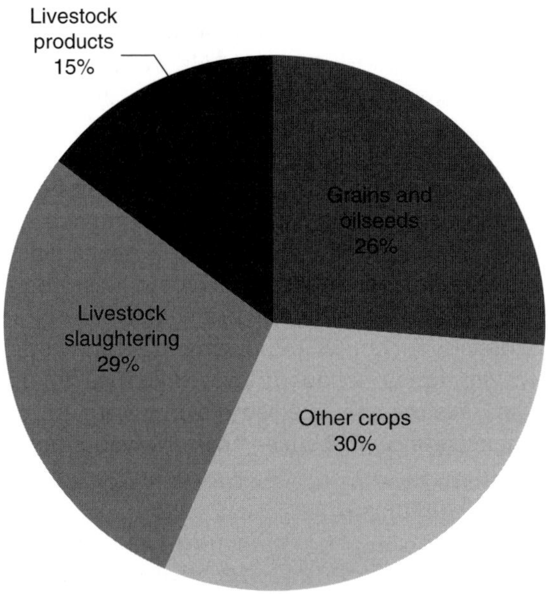

Fig. 2.2. Composition of gross value of Australian farm production 2010–2011 (based on ABARES, 2011, p. 16).

Over the past five decades, the share of contribution to the gross value of farm production from crops has been increasing, while that of livestock has been declining. In 1965–1966, livestock contributed 64%, but the contribution was down to 44% by 2010–2011. Since the mid-1990s, livestock contribution has been below 50%, except in 2002–2003. Hence, the importance of crop production has increased in Australia over the past two decades.

Crops are placed into two broad categories: winter and summer crops. Winter crops include wheat, barley, oats, triticale, lentils, lupins, chickpeas, faba beans, field peas, canola, linseed and safflower. Summer crops include rice, soybean, cotton, sorghum, maize, sunflower, groundnuts, mung beans and navy beans. Winter crops account for around 85–90% of total crop output. Of the states, WA is the largest winter crop producer, followed by NSW. QLD and NSW are the two major summer crop producers. The production of summer crops in all other states is minimal, being less than 1% of the national total (ABARES, 2011, p. 26).

In the past five decades, crop output increase in Australia has been very impressive. Table 2.5 shows the changes in outputs of major crops since 1965. Crop production is heavily weather dependent. Weather conditions affect crop yield and area sown or area harvested, and subsequently total output. The quality of the crops harvested is also affected by weather conditions. In some cases, crops harvested may be of only feed grain grade, reducing farm income. As has already been mentioned elsewhere, farming in Australia suffered from long-lasting and serious droughts in the first decade of the 2000s.

Weather conditions also affect livestock production. Feed availability from the pasture can be affected significantly by adverse weather conditions. For example, when severe droughts or floods hit, feed available from the paddock is limited and grains have to be supplemented to animals, increasing costs significantly. In the meantime, adverse weather may cause serious damage to crops. Reduced crop output affects the availability of feed grains and in turn affects livestock production, especially the operation of feedlots. Table 2.6 shows the changes in outputs of livestock products since 1965.

A brief note is provided below for several products where peculiar circumstances exist. Coarse grains have the second largest share in crop production after wheat, with around 30% of area sown to them. Major coarse grains include barley, oats, sorghum, maize and triticale, with barley having the largest share. What is interesting in regard to coarse grain production is that maize production in Australia is minimal. This seems to be because: (i) maize production requires a large amount of water; and (ii) the relative return is not attractive.

Paddy rice is one of the few crops for which irrigation is widely used. Dry land rice production is not common in Australia. Because of irrigation, yield is very high by world standards. In 2009–2010, national average yield reached the highest ever, 10.4 t/ha. However, in recent years, restrictions have been in place for water used in rice production. When there is a general shortage of water, rice production is restricted. This explains why in some recent years the area sown to rice has been so low (e.g. 2200 ha in 2007–2008 and 7200 ha in 2008–2009). Water availability will continue to affect Australia's rice production in a major way.

Dairying shows a typical example of productivity gain in Australian agriculture. Since 2000, the number of dairy cows has dropped from 2176 thousand head in 2000–2001 to 1604 thousand head in 2010–2011, a drop of 26.3%. However, the drop in total milk production was relatively small, being 13.7%, due to improvements in milk yield per cow. In 2000–2001, milk yield per cow was 4847 l and this increased to 5675 l in 2010–2011, an increase of 17.1%.

Sheep numbers reached a high in the late 1960s. From 1970, sheep numbers started to decline. The declining trend was arrested following the introduction of the wool minimum price scheme in 1974. Sheep numbers started to increase

Table 2.5. Output of major crops in Australia, 1965–2011 (based on various tables of ABARES, 2011).

		Unit	1965–1966	1970–1971	1980–1981	1990–1991	2000–2001	2010–2011
Coarse grains	Area sown	ha (thousands)	2,710	4,191	4,284	4,127	5,326	5,638
	Yield	t/ha	0.87	1.31	1.22	1.64	2.05	2.20
	Production	kt	2,371	5,474	5,209	6,766	10,914	12,389
	Exports	kt	478	2,218	2,764	3,310	4,629	5,337
	Exports/production	%	20.2	40.5	53.1	48.9	42.4	43.1
Barley	Area sown	ha (thousands)	930	2,000	2,451	2,555	3,454	3,740
	Yield	t/ha	1.02	1.18	1.09	1.61	1.95	2.18
	Production	kt	949	2,351	2,682	4,108	6,743	8,145
	Exports	kt	227	1,123	2,076	2,914	4,146	4,625
	Exports/production	%	23.9	47.8	77.4	70.9	61.5	56.8
Cotton	Area harvested	ha (thousands)	22.2	35.1	83.6	279.0	527.3	590.2
	Lint yield	t/ha	N/A	N/A	1.183	1.602	1.553	1.521
	Lint production	kt	N/A	N/A	99	447	819	898
	Exports	kt	N/A	7	59	340	834	505
	Exports/production	%			59.7	76.0	101.9	56.2
Oilseeds	Area sown	ha (thousands)	58	269	400	599	2,144	2,759
	Yield	t/ha	0.76	0.76	1.12	1.74	1.45	1.37
	Production	kt	44	204	450	1,040	3,098	3,782
	Exports	kt	0	9	6	187	2,194	1,728
	Exports/production	%	0.0	4.4	1.3	18.0	70.8	45.7

Rice	Area sown	ha (thousands)	26	41	104	89	177	75
	Yield	t/ha	7.0	7.4	7.3	8.8	9.3	9.7
	Production	kt	182	300	760	787	1,643	726
	Exports	kt	64	96	276	224	602	65
	Exports/production	%	35.2	32.0	36.3	28.4	36.6	9.0
Sugar	Area harvested	ha (thousands)	204	221	288	339	403	334
	Cane crushed	t/ha	14,382	17,645	23,976	25,200	28,117	27,443
	Sugar production	kt	1,924	2,448	3,227	3,407	4,162	3,610
	Exports	kt	N/A	N/A	N/A	2,649	3,087	2,514
	Exports/production	%				77.8	74.2	69.6
Wheat	Area sown	ha (thousands)	7,088	6,478	11,283	9,218	12,141	13,645
	Yield	t/ha	1.00	1.22	0.96	1.63	1.82	2.04
	Production	kt	7,067	7,890	10,856	15,066	22,108	27,891
	Exports	kt	4,755	9,049	9,614	12,002	16,142	18,639
	Exports/production	%	67.3	114.7	88.6	79.7	73.0	66.8

Table 2.6. Output of major livestock products in Australia, 1965–2011 (based on various tables of ABARES, 2011).

		Unit	1965–1966	1970–1971	1980–1981	1990–1991	2000–2001	2010–2011
Dairy products	Dairy cow numbers	thousands	3,094	2,745	1,819	1,682	2,176	1,604
	Milk yield/cow	l	2,236	2,641	2,882	3,807	4,847	5,675
	Milk production	Ml	6,919	7,249	5,243	6,403	10,547	9,102
	Butter production	kt	209	203	79	06	172	122
	Cheese production	kt	60	78	135	179	376	338
Beef and veal	Cattle numbers	million	18.81	23.73	25.17	25.37	27.72	28.81
	Slaughtering	million	6.56	5.73	8.83	8.25	8.73	8.27
	Average weight	kg	154	176	174	211	235	257
	Production	kt	1,013	1,011	1,534	1,738	2,053	2,129
	Exports	kt	N/A	N/A	883	1,052	1,329	1,358
	Exports/production	%			57.56	60.53	64.73	63.79
Mutton and lamb	Sheep numbers	million	170.6	171.7	131.4	166.6	110.9	74.3
	Slaughtering (mutton)	million			14.6	17.3	16.4	6.2
	Average weight	kg			19.6	20.7	21.1	22.6
	Production	kt			286.4	358	345.7	139.1
	Exports	kt			229.8	212.8	235.5	128.6
	Exports/production	%			80.24	59.44	68.12	92.45
	Slaughtering (lamb)	million			15.7	16.5	18.5	18.6
	Average weight	kg			16.6	17.5	19.9	21.6
	Production	kt			260.6	289	368.2	401.9
	Exports	kt			43.1	43.5	125.2	186.4
	Exports/production	%			16.54	15.05	34.00	46.38
Wool	Wool production (greasy)	kt	754.2	890	701.2	989.2	645.1	429.1
	Clean yield	%	58.6	59	62.8	65.6	65.8	64.9
	Closing stock	kt		73.9	30.3	820.1	141.4	N/A
	Exports	kt				715	850	443

again, from a low of 127.5 million in 1977–1978 and reached another high of 173.8 million in 1989–1990. Increased wool production, coupled with reduced global demand for wool due to increasing competition from synthetic fibres, resulted in an oversupply of wool. From 1990, sheep numbers again started to drop. By 2010–2011, Australia's sheep number dropped to a low of 74.3 million, less than half of the number in the peak time. Wool production has also dropped, from 1031 thousand t in 1989–1990 to a little over 400 thousand t currently (ABARES, 2011, p. 224). Despite the reduction in wool output, Australia still produces a large share of the world's supply of fine wool.

Sheep meat output has also fluctuated. However, lamb output has shown an increasing trend as a result of recent higher prices. The demand for lamb from both domestic and overseas consumers has been on the increase in recent years. Some wool producers have shifted to produce more lamb. Changing consumer preference for woollen products may lead to further reduced demand for wool, leading to possible further decline in wool production.

Pork and poultry meat each account for some 10% of total Australian meat production. Both Australia's pork and poultry meat production are chiefly for domestic consumption. In recent years, per capita consumption of both of these two meats has been increasing. There is further scope for pork and poultry meat production to expand for domestic markets.

2.3.7 Outputs of fisheries and forestry

Together, about 14% of rural GDP is produced by the fisheries and forestry sectors. Fishing historically has been a relatively small industry in Australia, largely because of plentiful supplies of domestically produced meat.[3] However, the value of Australian fisheries production has increased, largely as a result of strong growth in exports of prawns, lobster, abalone, salmon and tuna to Asian countries. Australian fishery production has actually dropped to some extent during the 2000s, from 279 thousand t in 2004–2005 to 241 thousand t in 2009–2010. However, the gross value rose from AUS$2086m to AUS$2185m, a 5% increase. This rise in production value reflects the trend towards the production of high-value species.

Australian fisheries operate in one of the world's cleanest environments. Australia spans a number of climatic zones and encompasses aquatic environments that can be divided broadly into open ocean, benthic reef, estuarine and inshore areas and freshwater habitats. Australia contains one of the most diverse marine faunas in the world, because of its geographical isolation from other continents and wide range of habitat types that encompass tropical to sub-Antarctic waters. The diversity and cleanliness of aquatic environments means that Australia is able to supply a vast array of delicious seafood products to the world.

Commercial fishing is one of the most valuable Australian rural industries. About 600 marine and freshwater seafood species are caught and sold in Australia for local and overseas consumption. Australia is one of the largest producers of abalone and rock lobster, and Australian South Sea pearls are considered the most valuable in the world.

Australia has the world's third largest fishing zone, covering 11 million km^2 and extending up to 200 nautical miles out to sea. Despite this impressive size, Australian waters tend not to be as productive as those in many regions and Australia ranks only 52nd in the world in terms of volume of fish landed.

The value of 'wild caught' seafood still dominates the Australian fishing industry. Recently, however, aquaculture production has gained momentum, being one of Australia's fastest growing rural industries. Currently, more than 60 species are being farmed, including pearl oysters, edible oysters, salmonoids, southern bluefin tuna, mussels, prawns, abalone, barramundi, yellowtail kingfish and freshwater finfish. The rise in the value of aquaculture in percentage terms indicates a longer-term trend, suggesting the sector will provide the major impetus for medium- to long-term growth in the value of Australia's seafood production.

Much of Australia's original forest cover has been cleared for agriculture. Today, only 19.42% of Australia's total land area is forested, most of which is woodland forest (ABARES, 2011, p. 119). Plantation areas (mainly conifers), though having expanded rapidly over the past few decades, are still small and cover only 0.26% of total land area. Hence, forestry operations in Australia are still mainly dependent on native forests. Forestry is an important industry in TAS and parts of coastal VIC and NSW.

Australia both exports and imports forestry products. It used to be a net importer of roundwood until the mid-1990s. Since then, Australia has been a net exporter of roundwood. In 2009–2010, Australia produced 25.6 million m^3 of roundwood, of which 20.9 million m^3 were consumed domestically. Woodchips account for a large portion of the value of Australia's exports of forest products, being about 40% on average during the recent decade, followed by paper and paperboard (28%) (ABARES, 2011, p. 123).

2.3.8 Agricultural terms of trade

Agricultural terms of trade are defined as the ratio of the index of prices received by farmers to the index of prices paid by farmers. From the middle of the 20th century, agricultural terms of trade showed a declining trend. This trend continued until the end of the 20th century (see Fig. 2.3). In the first decade of the 21st century farmers' terms of trade were still declining, but more slowly. Wonder and Fisher (1990) attribute this long-term declining trend in agricultural terms of trade to worldwide increases in the technical efficiency of agricultural production, as well as to the significant assistance offered to producers in the EU, Japan and the USA.

Examining Fig. 2.3, there were times when farmers' terms of trade improved. In the early 1970s, the terms of trade were improved when the prices received for beef and wheat recorded significant increases (Wonder and Fisher, 1990). Then, from 1978–1979 to 1982–1983, the prices received for beef, sheep meat and wheat again rose significantly. During the same period, the prices paid by farmers initially rose at a much slower pace but later increased faster, especially in 1981–1982 and 1982–1983, leading to the decline in the terms of trade. Farmers' terms of trade again improved between 1987–1988

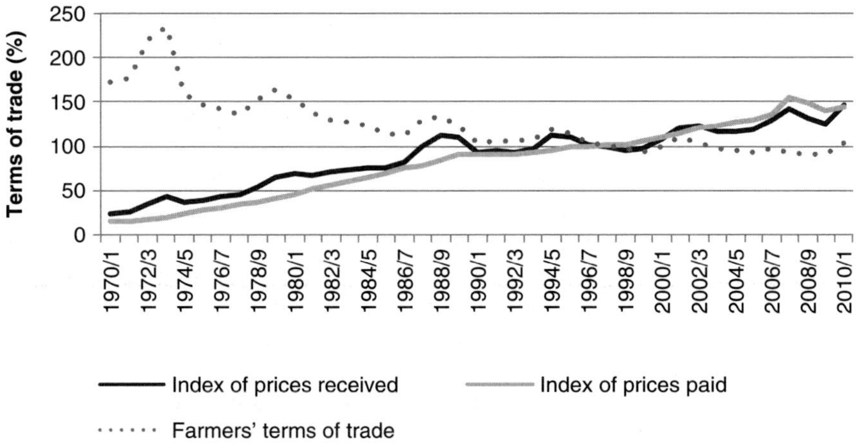

Fig. 2.3. Agricultural terms of trade in Australia (based on ABARES, 2011, p. 14).
Note: Agricultural terms of trade is the ratio of the index of prices received by farmers to the index of prices paid by farmers, 1997/8 = 100.

and 1988–1989 as a result of strong increases in prices for crops such as wheat, barley, wine grapes and potatoes. Since 1990–1991, there have been no dramatic improvements or deteriorations in terms of trade as happened around 1979–1980 and 1988–1989, but with a few noticeable improvements, for example, in 1994–1995, 2001–2002 and 2010–2011. These improvements were as a result of a faster increase in the prices received by farmers, led by wheat, barley, oats, cotton, beef and wool.

Despite the occasional improvements, overall, farmers' terms of trade have been declining since the early 1970s, as shown in Fig. 2.3. On the other hand, the level of agricultural output has been increasing in real terms over the same period (ABARES, 2011, p. 14). A steady improvement in productivity is the key reason why agricultural output has been able to grow while the terms of trade have been declining.

2.3.9 Land and water use

Land used for farming purposes has declined over the past four decades (Table 2.7). Though water use statistics for earlier years are not available, the data for 2005–2006 until 2009–2010 show that water use for agriculture is declining.[4] Earlier, we pointed out that fertilizer use in farming was also declining. However, agricultural outputs have been increasing over the past five decades (Tables 2.5 and 2.6). This suggests that Australian farmers are producing more with less.

2.3.10 Rural contribution to GDP and exports

GDP produced by the rural sector has increased in size. In 1974–1975, it was AUS$11,832m (in nominal value). By 2010–2011, it had expanded to

AUS$34,040m. In terms of its contribution to the total GDP, it has remained below 3% since 1974–1975. The total value of exports from the rural sector has also expanded significantly, from AUS$3972m in 1974–1975 to AUS$36,252m in 2010–2011 (Table 2.8). However, its share-out of national total exports has continued to slide. It was 60% in 1965–1966. In 1974–1975, it slid to 39%. By 2010–2011, this share dropped further, to a little over 12%.

Despite the decline in relative importance, agriculture has always been regarded as playing a very important role in the Australian national economy. It is widely held that the performance of the rural sector has a significant impact on incomes in the rest of the economy (Williams, 1967; Stoeckl and Miller, 1982;

Table 2.7. Farm use of land and water in Australia (ABARES, 2011, p. 22 and p. 205).

	Farm use of land (m ha)			Farm water use (Gl)		
	Wheat	Other crops	Total	Irrigation	Other uses	Total
1970–1971	6.5	4.7	497.2			
1980–1981	11.3	5.2	495.4			
1990–1991	9.2	6.6	462.8			
2000–2001	12.1	10.4	455.7			
2001–2002	11.5	10.5	447.0			
2002–2003	11.2	10.2	439.5			
2003–2004	13.1	10.7	440.1			
2004–2005	13.4	11.0	445.1			
2005–2006	12.4	10.4	434.9	10,737	951	11,689
2006–2007	11.8	9.9	425.4	7,636	885	8,521
2007–2008	12.6	11.2	417.3	6,285	704	6,989
2008–2009	13.5	11.1	409.0	6,501	785	7,286
2009–2010	13.9	10.5	398.6	6,596	763	7,359
2010–2011	13.6	10.8	N/A	N/A	N/A	N/A

Table 2.8. Value of rural exports (nominal), 1965–2011 (based on ABARES, 2011, p. 3).

	Value of rural exports (AUS$m)				Percent of total goods and services			
	Farm	Forest	Fisheries	Total rural	Farm	Forest	Fisheries	Total rural
1965–1966	1,849	15	24	1,888	58.7	0.5	0.8	60.0
1970–1971	2,103	25	56	2,184	41.2	0.5	1.1	42.8
1975–1976	4,267	84	80	4,431	37.9	0.7	0.7	39.4
1980–1981	8,179	238	230	8,647	36.1	1.0	1.0	38.1
1985–1986	6,332	348	482	7,162	16.2	0.9	1.2	18.3
1990–1991	13,076	706	831	14,612	19.7	1.1	1.2	22.0
1995–1996	20,598	1,087	1,328	23,013	20.6	1.1	1.3	23.1
2000–2001	30,078	1,846	2,169	34,093	19.3	1.2	1.4	21.9
2005–2006	27,815	2,140	1,547	31,501	14.2	1.1	0.8	16.1
2010–2011	32,529	2,474	1,249	36,252	10.9	0.8	0.4	12.2

O'Mara, 1987; Wonder and Fisher, 1990). Changes in the value of rural production are transmitted in the economy through various linkages. Major such linkages include: consumption, investment, input use, expenditure switching, competition for scarce resources and the exchange rate and the level of wages (Crofts *et al.*, 1988; Wonder and Fisher, 1990). The performance in Australian agriculture will continue to have a major impact on the whole Australian economy.

2.4 Concluding Comments

Despite the harsh natural environment, competitive agricultural markets and declining terms of trade, farmers in Australia have been successful in advancing Australian agriculture, making it one of the most efficient and modern agricultures in the world. Many factors will have contributed to this success. In this book, it is argued that, among various factors, the major ones responsible for having shaped Australian agriculture as it is today include:

- institutional arrangements conducive to agricultural development;
- protecting farmers' interests;
- deregulating the agricultural industry;
- fostering farmer entrepreneurship;
- proactive agricultural R&D investment;
- encouraging agricultural innovations; and
- promoting sustainable farming.

Chapters 3–9 in the next part of this book elaborate on each of these factors to show how they have contributed to Australian agricultural success.

Notes

[1]Unless indicated otherwise, much of the content in this section is based on *Year Book Australia 2008* (ABS, 2008a).
[2]This figure is different from the one in Table 2.1 for 2009–2010 because the latter was the figure for the end of June 2010.
[3]Part of the discussion on Australian fisheries is based on the ABS (2008a, pp. 509–520).
[4]It is not certain whether this trend will continue. The long drought in the first decade of the 2000s might have contributed to this decline in water use. None the less, the ABARES data (2011) showed that outputs of most farm products grew, while water use declined.

References

ABARES [Australian Bureau of Agricultural and Resource Economics and Sciences] (2011) *Agricultural Commodity Statistics*. Australian Government, Canberra.
ABS [Australian Bureau of Statistics] (2008a) *Year Book Australia 2008*. Australian Government, Canberra.
ABS (2008b) *Natural Resource Management on Australian Farms, 2006–07*. Cat. No 4620.0. Australian Government, Canberra.

ABS (2011) *Agricultural Land and Water Ownership Survey, December 2010*. Cat. No 71270DO001_201012. Australian Government, Canberra.

Ashton, P. (1988) *Australian Farming Through 200 Years – 200 Years in Pictures*. Kangaroo Press, Kenthurst, NSW, Australia.

Clark, M. (1992) *A Short History of Australia*. Penguin Books, Ringwood, VIC, Australia.

Crofts, B., Harris, M. and O'Mara, P. (1988) Variation in farm output and its effect on the non-farm sector. *Quarterly Review of the Rural Economy* 10, 249–254.

Davidson, B.R. (1981) *European Farming in Australia*. Elsevier, Amsterdam.

Dwyer, J. and O'Mara, P. (1988) Measuring Australia's competitiveness. *Quarterly Review of the Rural Economy* 10, 54–59.

Gruen, F. (1990) Economic development and agriculture since 1945. In: Williams, D.B. (ed.) *Agriculture in the Australian Economy*, 3rd edn. Sydney University Press, Sydney, Australia, pp. 19–26.

Henzell, T. (2007) *Australian Agriculture: Its History and Challenges*. CSIRO Publishing, Collingwood, VIC, Australia.

Mullen, J.D. (2001) *An Economic Perspective on Land Degradation Issues*. Economic Research Report No 9, NSW Agriculture, Orange, Australia (http://www.dpi.nsw.gov.au/research/economics-research/reports/err09, accessed 12 June 2012).

O'Mara, L.P. (1987) The contribution of the farm sector to annual variations in gross domestic product in Australia. *Economic Record* 63, 255–269.

Shaw, A.G.L. (1982) History and development of Australian agriculture. In: Williams, D.B. (ed.) *Agriculture in the Australian Economy*, 2nd edn. Sydney University Press, Sydney, Australia, pp. 1–28.

Shaw, A.G.L. (1990) Colonial settlement 1788–1945. In: Williams, D.B. (ed.) *Agriculture in the Australian Economy*, 3rd edn. Sydney University Press, Sydney, Australia, pp. 1–18.

Stoeckl, A. and Miller, G. (1982) Agriculture in the economy. In: Williams, D.B. (ed.) *Agriculture in the Australian Economy*, 2nd edn. Sydney University Press, Sydney, Australia, pp. 166–185.

Williams, D.B. (ed.) (1967) *Agriculture in the Australian Economy*. Sydney University Press, Sydney, Australia.

Wonder, B. and Fisher, B. (1990) Agriculture in the economy. In: Williams, D.B. (ed.) *Agriculture in the Australian Economy*, 3rd edn. Sydney University Press, Sydney, Australia, pp. 50–67.

II What Has Made Australian Agriculture Successful

3 Getting the Institutions Right

Over the 20th century, Australia established various institutional arrangements that had an impact on agricultural development. Such institutional arrangements were, and still are, fundamental to Australia's agricultural success. This chapter explains these important institutional arrangements and demonstrates how they have facilitated or affected Australia's agricultural development.

3.1 The Constitutional Framework

The current Australian Constitution came into effect in 1901 when federation occurred. There are two sections in the Constitution that have significant impacts on agricultural development, namely, Section 51 and Section 92.

The Australian Constitution defines the boundaries of law-making powers between the Commonwealth and the states/territories. Section 51 of the Constitution lists all the powers rendered to the Commonwealth government. Powers not granted to the Commonwealth in this section are regarded as residual powers. All residual powers remain with the state/territory governments. According to Section 51 of the Constitution, the Commonwealth is not given direct responsibility for agriculture, except for trade and commerce with other countries; state/territory governments are, on the other hand, responsible for many aspects of agriculture under their jurisdiction. They have sovereign powers in matters affecting their rural industries, such as the regulation of agricultural production and marketing within their borders, land tenure, land use and water supply.

However, Section 92 stipulates that trade between states should be free and hence this limits the powers of states to act independently. According to Section 92, 'on the imposition of uniform duties of customs, trade, commerce, and intercourse among the States, whether by means of internal carriage or ocean navigation, shall be absolutely free'.

The constitutional division of powers between the federal and state governments and the Constitution's insistence on freedom of interstate trade have had the most pervasive and powerful effects on the development of Australian agriculture. There have been outcries about the consequences caused by these two constitutional clauses (see, for example, Lewis, 1967; Campbell, 1980, pp. 90–94; Lloyd, 1982). The often-cited major 'consequences' of Section 51 are:

- impediments to national regulatory activities, research problems of national significance, national statutory marketing schemes, national schemes of price support and other forms of federal assistance;
- different types of legislation in different states dealing with the same issue; and
- enormous needs for coordination between federal and state activities.

These consequences did indeed cause much grief to farmers, farm suppliers and federal and state agricultural administrators prior to the 1990s. Since the early 1990s, the deregulation of markets across various industries has meant that the conflicts caused by Section 92 have largely disappeared (see Chapter 5). The need for coordination and collaboration between federal and state governments and between state governments themselves caused by Section 51, however, still exists. What follows highlights briefly the approaches used for such coordination and collaboration.

As a result of the division of power between the federal and state governments in agricultural administration, departments of agriculture (or department of primary industries, DPI) at both the state and federal levels work, to a great extent, independently. There is no direct supervisory relationship between federal and state departments of agriculture. None the less, cooperation and collaboration, as noted earlier, are needed and, in some cases, are indispensable. For example, in the case of preventing or eradicating pests and diseases and marketing produce internationally.

There had been the desire to coordinate matters concerning agricultural development at the national level. Prior to 1934, matters of national importance and relevance were handled by correspondence or at ad hoc conferences of state/territory and, at times, federal government ministers. In 1935, the Australian Agricultural Council was established, which helped coordination at the national level (Jessup and Dun, 1982).

From its inception, New Zealand attended meetings of the Australian Agricultural Council as an observer. In 1991, New Zealand was invited to participate as a full member of the council. Following acceptance of full membership, it was agreed (at the February 1992 meeting) to change the name to the Agricultural Council of Australia and New Zealand.

In October 1992, the Agricultural Council of Australia and New Zealand was replaced by the Agriculture and Resource Management Council of Australia and New Zealand (ARMCANZ), under which the then Australian Soil Conservation Council (ASCC) and the Australian Water Resources Council (AWRC) were brought. In June 1993, the Council of Australian Governments (COAG) reviewed and rationalized some other ministerial councils, and as a result, added the responsibilities of the Rural Adjustment Scheme (RAS) Ministers' Meeting to those of the newly established ARMCANZ. ARMCANZ had its first meeting in

Alice Springs in July 1993. It was the peak government forum for consultation, coordination and, where appropriate, integration of action by governments on agriculture, land and rural and urban water issues.

Over the years, to facilitate coordination and collaboration between federal and state governments to deal with a wide range of other issues, such as forestry and fisheries matters, several other ministerial councils were also established. For example, the Australia New Zealand Environment and Conservation Council (ANZECC) was formed in July 1991 by the amalgamation of the former Australian Environment Council (AEC, established in 1972) and the former Council of Nature Conservation Ministers (CONCOM, established in 1974). New Zealand was admitted to full membership of the AEC and CONCOM in July 1989. Another council that was also created in 1993 was the Ministerial Council on Forestry, Fisheries and Aquaculture (MCFFA). This council was created from the amalgamation of the Australian Forestry Council with the Australian and New Zealand Fisheries and Aquaculture Council. This Council comprised the Australian/state/territory and New Zealand government ministers responsible for forestry, fisheries and aquaculture.

During 1999–2000, the debate on the impact of natural resource degradation in Australia began in earnest. Among other things, this resulted in the establishment of a new Natural Resources Management Ministerial Council (NRMMC). As a consequence of this, all natural resource management (NRM) issues previously dealt with by existing councils such as ANZECC, ARMCANZ and MCFFA were transferred to the new NRMMC. The residual 'industry-related' issues of these latter two councils were brought together under a new Primary Industries Ministerial Council (PIMC) (SCoPI, 2012).

The NRMMC and PIMC came into existence in 2001. The particular importance of these two ministerial councils in the Australian context is to facilitate the implementation, nationally, of plans and proposals that would not otherwise be possible because of the limitations imposed by the division of constitutional powers between Australian and state/territory governments.

The NRMMC consists of the Australian/state/territory and New Zealand government ministers responsible for primary industries, natural resources, environment and water policy. It is the peak government forum for consultation, coordination and, where appropriate, integration of action by governments on NRM issues. Its objective is to promote the conservation and sustainable use of Australia's natural resources.

The PIMC consists of the Australian/state/territory and New Zealand government ministers responsible for agriculture, food, fibre, forestry, fisheries and aquaculture industries/production and rural adjustment policy. It is the peak government forum for consultation, coordination and, where appropriate, integration of action by governments on primary industry issues. Its objective is to develop and promote sustainable, innovative and profitable agriculture, fisheries/aquaculture and food and forestry industries.

These two councils have meetings about twice a year. All the communiqués from their first meeting are available for public access from their website (http://www.mincos.gov.au/media_releases/media_archives#nrmmc). All important issues dealt with at each of the meetings are available for perusal in these communiqués.

In February 2011, the COAG announced that a standing council on primary industries (the Standing Council on Primary Industries) and another on environment and water (the Standing Council on Environment and Water) would be established. These two councils would replace the NRMMC and PIMC. The NRMMC ceased operation in June 2011 and the PIMC in September 2011. The Standing Council on Primary Industries (SCoPI) was launched in September 2011.

SCoPI became the peak forum to pursue and monitor priority issues of national significance affecting Australia's primary production sectors, which required a sustained and collaborative effort across jurisdictions, and to address key areas of shared Commonwealth, state and territory responsibility and funding for Australia's primary production sectors (SCoPI, 2012).

Given the inconveniences caused by the two constitutional clauses, one might have wondered why Australians did not make amendments to them. They tried, but making any amendments to the Constitution is not easy. Section 128 of the Constitution provides that constitutional amendments must be approved by a referendum. Successful amendment requires:

- an absolute majority in both houses of the federal parliament;
- the approval in a referendum of the proposed amendment by a majority of electors nationwide; and
- a majority in a majority of states.

Getting the double majority is not a result that can be achieved easily. To date, 44 proposals to amend the Constitution have been voted on at referenda, of which only eight have been approved. Of these 44 proposals, several were about obtaining increased federal control over the marketing of primary products. None of them was approved.

Is the division of powers between the federal and state governments and the lack of federal control over the marketing of primary products all bad? Not necessarily. True, the drafters of the Constitution at the end of the 19th century could not foresee that government intervention in agriculture would have increased significantly in the 20th century. The same is also true that those who proposed to amend the Constitution in the early 20th century to enable greater federal control over agricultural activities could not foresee that many such controls would no longer be needed by the end of the 20th century.

One could argue that centralization of power over agriculture was needed. In fact, during the Second World War, federal control over agricultural production and marketing was achieved by virtue of the defence powers of the Commonwealth government. However, the centralization of power over agriculture in this period did not seem to be resoundingly successful as far as the various primary industries were concerned, though it did cut down on the need for federal–state consultation and collaboration (Campbell, 1980, p. 90). After the war and the failure of the 1946 referendum, the powers acquired temporarily by the federal government were restored to the states.

On the other hand, the division of powers over agriculture between the federal and state governments has several advantages:

1. It enables different states to have the flexibility to deal with and focus on issues peculiar to their own states, providing interstate trade remains free.
2. Different approaches in different states tackling the same issue can lead to very innovative solutions that other states can learn from.
3. The need to bring senior officials of federal and state governments together to discuss and negotiate on issues on which they need to collaborate, though cumbersome, allows different views to be heard and also can lead to creative approaches to dealing with such issues.
4. States' autonomy in administering their agriculture helps them to utilize the comparative advantages of their states best to develop their agricultural industries.

Between the federal and state/territory governments, disagreements or conflicts on agriculture-related issues do occur sometimes. John Kerin (former Federal Minister for Agriculture under Labor, 1983–1991) and John Anderson (former Federal Minister for Agriculture under the Coalition, 1996–1998) explain how such disagreements or conflicts are resolved. Their answers are very similar. That is, the federal government initiates a process of persuasion and negotiation. All parties are encouraged to put forward good options. Final decisions are made by following the principle that what is to be done should be good for the whole country. Both of them admit that negotiation can be a slow process, but it works. They also point out that should negotiations break down and issues remain unsolved but the federal government really believes something has to be done, the federal government may use money allocation through 'state payment' to 'convince' the states to carry out the necessary changes or reforms (J. Kerin, Canberra, 16 November 2010; J. Anderson, Sydney, 15 December 2010, personal communication).

3.2 Land Tenure

In addition to the constitutional arrangements, the other legal framework that vitally affects the performance of agricultural industries is land tenure. Campbell (1980, p. 96) points out that, whether established by custom and tradition or codified in the law, regulations governing the ownership and use of land have a vital effect on the profitability of farm businesses, on the distribution of income in the rural community and on the degree to which the land is conserved or exploited.

There are two major types of tenure for rural land in Australia today: leasehold and freehold. Leasehold is the major type of land tenure. This is attributable simply to the fact that, in Australia, the majority of land (72%) is Crown (or government) owned. Privately owned land accounts for only 15%. The remaining 13% is owned by Aboriginal communities. Table 3.1 gives the breakdown of land ownership by government, Aboriginals and private owners in various states.

Australia is believed to be unique among developed countries in that a high proportion of its land is in public ownership. Over the past 220 years, only 15% of the total area of the country has been alienated (i.e. privately owned). The alienated lands, for the most part, are located in the earlier settled areas, as shown in

Table 3.1. Land ownership in Australia (based on Cooray, 1995; the table was compiled by Cooray from a number of sources).

	Government		Aboriginals		Private/freehold		Total
	m ha	%	m ha	%	m ha	%	m ha
WA	211	84	22	9	19	7	252
QLD	135	78	3	2	35	20	173
NT	80	60	55	40	0	0	135
SA	65	66	18	19	15	15	98
NSW	1	64	3	3	26	33	80
VIC	9	38	–	–	14	62	23
TAS	4	62	–	–	3	38	7
ACT	0.2	100	–	–	0	0	–
Total	555	72	101	13	112	15	768

Table 3.1. In VIC, a little over 60% of its lands are in private ownership. In NSW, about one-third of the occupied land is owned privately. Some two-fifths of the land in TAS is in the hands of private owners. In other states, i.e. WA, SA and QLD, less than 20% of the land is privately owned. In both the ACT and NT, private ownership of rural lands is zero.

The types of tenures for rural land in different states are broadly similar. However, terms and conditions applying to particular tenures tend to vary from state to state. The two major types of land tenures, leasehold land and freehold land, are explained briefly below using the information provided by the WA government (DIA, 2010) and the QLD government (DERM, 2011).

3.2.1 Leasehold

Leasehold land is Crown land over which the Crown (or the government) has granted an interest. That interest is a lease (or in other words the Crown has rented its land to an individual or other legal entity). The person or organization that was granted the leasehold can hold, occupy and use the land and in return pay a rent.

The grant of leasehold carries with it certain conditions or requirements. If the requirements of the leasehold are not met, then the Crown may take back the land. In most cases, the leasehold is for a set time and for a specific purpose. The leasehold will expire on a certain date, but can be renewed subject to any conditions regarding management and development of the land.

Leasehold land can be sold with the approval of the relevant government minister.[1] Depending on the leasehold's purpose and the length of the lease, leasehold land can be used sometimes as security to raise loans from banks and other financial institutions (but again with the approval of the relevant government minister).

Leases in perpetuity have no expiry date: they go on forever. For most practical purposes, land held under perpetual lease is virtually the same as freehold land (see Section 3.2.2 below). However, perpetual leases still have conditions or requirements attached to them like any other leases and the approval of the relevant government minister is required for the transfer of leases in perpetuity.

The lessees have certain rights. If they comply with the lease conditions, they are then entitled to use the land without any interference from the government. Also, they own all the improvements on the land. The government only owns the unimproved land. When the lessees sell the lease, they will realize the value added to the land through improvements.

There are different types of lease that are issued by a state department, often the department of agriculture, or by a reserve's management body (for the Crown). Pastoral leases are leases for grazing stock and all purposes connected with that. General leases are leases for some particular purpose, for example, for the purpose of the use and benefit of Aboriginal people. Reserve leases are leases for the purpose of reserve.

Pastoral leases are a particular type of leasehold that allows Crown land to be used for grazing stock. The leases are granted by state departments of agriculture (for the Crown). All pastoral leases will expire at a certain time, but they can be renewed if the lessee has met the conditions regarding the management and development of the land.

Major conditions related to pastoral leases include: rent has to be paid; the land must be managed in an environmentally sound way; the leaseholder cannot sell the lease or use the lease as security to raise money without approval from the relevant government minister. Also, the Crown has the right to the compulsory purchase (in other words the Crown can buy the lease from the leaseholder even if he or she does not want to sell it) of any part of the lease for purposes like state improvement or settlement.

If a pastoral leaseholder fails to meet the lease conditions, the lease can be taken away. Before that happens, a default notice (describing what has not been done, or has been done wrongly) is issued. The notice requires the leaseholder to comply with the lease conditions (in other words to put right what they have been doing wrong). If the leaseholder does not do what the default notice demands, then a 'Notice of Intention to Forfeit' is issued. This notice means that the lease will be taken away.

Apart from having the lease taken away, fines may also apply. In many cases, a permit is needed from a state authority to do particular things on the land under a pastoral lease. The lack of a permit if it is required usually results in a fine for the person who holds the lease. For example, various amounts of fines may be applied, depending on the violation:

- when the land is used for purposes that are not pastoral without having a permit;
- when the land is cleared (trees, shrubs and other plants removed) without a permit, the fine will be AUS$10,000;
- when plants that are not natural to the land are planted and sold without a permit;

- when other people's stock is kept on the land in return for a fee without a permit;
- if the person who holds the lease does not put in a yearly stock and improvement notice, or if the person deliberately provides false information on that notice;
- when the lease is held by a company, if any shares in the company are transferred without approval; and/or
- if the person who holds the lease receives a default notice and does not comply with it.

Rent will have to be paid to the Crown. Lessees may also need to pay rates to local governments and, if applicable, charges for vermin control.

3.2.2 Freehold

Freehold land is land over which the Crown has granted an interest. The freehold interest is the least restricted interest in land and is usually known as 'ownership' of land. Unlike leasehold, the land is no longer called Crown land after the freehold interest has been granted. The Crown cannot put conditions on the use of the land as it can with leasehold. There may be some restrictions on how the land can be used in order to conserve resources.

Freehold gives the owner of that interest the exclusive right to the land for an indefinite period. The owner of the freehold can sell the land to anyone else. The owner may also lease the land to someone else on whatever conditions they like.

When the Crown grants the freehold interest, it keeps the right to buy the land back compulsorily for public works such as roads, railways, bridges, schools, hospitals or other purposes to benefit the state. When freehold land is needed for this type of use, the Crown first tries to buy the land by agreement with the owner. If that fails, then the Crown will buy the land compulsorily and pay compensation to the owner. Generally, the compensation is very much in favour of the owner; the drafters of the Constitution have made it very difficult for the government to buy back the land when the owner is not in agreement with selling the land back.

Although owners of freehold land do not need to pay rent to the Crown, they still need to pay rates to the local government and any charges for vermin control, the same as those owners of leasehold land. As shown in Table 3.1, states settled earlier, such as VIC, TAS and NSW, have a much higher proportion of freehold land. It is also believed that freehold land is generally better land (B. Malcolm, University of Melbourne, 7 October 2010, personal communication).

3.2.3 Other types of land tenure

Leasing of land from private individuals

This occurs but is not common in Australia. Most owners of freehold land are running their own family farms. Limited freehold land would be available for leasing out. The leasing of leasehold land must go through the approval procedures, as indicated earlier, and it can take time and be cumbersome.

Share-farming

This is practised in Australia but tends to be confined to certain industries: that is, the dairy, wheat and potato industries. There is considerable variation in the respective obligations and the division of the output between the landlord and the share-farmer (Campbell, 1980, pp. 98–99).

3.2.4 Is the high proportion of public land ownership an issue?

The high level of public land ownership does not seem to cause much concern among Australian agricultural economists, policy makers, farmers and the general public. Indeed, discussions on public land ownership are quite rare in academia and the media. None the less, arguments about the very high proportion of public land ownership do exist: see, for example, Campbell (1980) and Cooray (1995).

Cooray (1995) argues that public ownership does not give farmers high security and does not encourage land improvements. In reality, however, 'security' does not seem to worry farmers. The simple reason is that the landowner, or the government, gives assurance to the lessee and the lessee's right to farm the land is protected by law. In addition, the lease is generally long and many of them are perpetual (perpetual lease land has no practical difference from freehold land, according to B. Malcolm (University of Melbourne, 7 October 2010, personal communication)). Because of the protection by law for farmers to farm the land, improving the land is a willing act many farmers carry out. Indeed, each year, Australian farmers spend billions of dollars on improving and protecting their land resources (see Chapter 9 for more details). The law also stipulates that all the improvements on the land are owned by the lessees. Farmers will be able to realize the value added to the land through improvements. As such, there is no disincentive for farmers not to improve the leased land.

Another issue related to the high level of public land ownership is the need for government approval of any lease of leasehold land. The approval may, on occasions, slow down the process of leasing land from private individuals. Campbell (1980, p. 98) believed this was a definite disability from the standpoint of facilitating structural changes on farms over time. He argued that farm businesses could be enlarged much more readily if leasing was made easier for farm operators. It can be the case that the leasing process may be slowed down by waiting for the necessary approval. However, it should not be overlooked that the checking and verification of the land conditions and the subsequent approval or disapproval is an important part of land resource protection efforts.

In sum, there is no evidence that suggests that the high level of public land ownership is a drag on the advancement of Australian agriculture. With law protection, farmers work on the leased land virtually the same as if they were working on their own land. The fact that any improvements to the land belong to the lessee encourages farmers to improve the land. Further, the lease arrangements actually help the Australian public, through the government, to keep land use under check, contributing to sustainable use of the land's resources.

3.3 Responsible Government, the Public Service and the Department of Agriculture

The Australian political system is an expression of two major principles: federalism and responsible parliamentary government. Responsible government involves a system of rules by which, once elections are held, governments are created, operate and are held responsible to the parliament and, indirectly, to the people who elected the parliament. Farmers are voters. Australian farmers also have their own organizations that can have dialogue with the governments (see Section 4.2 in the next chapter). Consequently, governments in Australia have to respond to farmers' needs and be responsible to them. Otherwise, a government may suffer from losing its parliamentary members in the next election as a result of reduced support from farmer voters.

Ministers for agriculture are usually members of government cabinets. These senior ministers are supported by departments of agriculture (or primary industries). State departments of agriculture have a long history because colonial and state governments have always had responsibility for agriculture. After the federation in 1901, these departments retained the powers and functions they had prior to 1901, except in respect of foreign trade. Departments such as agriculture are part of the public service through which the Australian governments make and administer their policies. In Australia, public servants are, by tradition, neutral, anonymous and apolitical servants of the state. The public servants in a department are responsible to their minister and for particular policies.

State departments of agriculture are responsible for industry regulations, agricultural research, development and extension (RD&E), advisory services and domestic marketing, and so on. The Commonwealth department of agriculture (or, sometimes, primary industries) is a relative newcomer and only emerged as a separate department with a minister of its own in the 1950s. By this time, the Commonwealth government developed a considerable interest in agricultural politics, particularly since the Second World War.

State departments of agriculture have long held a reputation for fiercely representing and defending the interests of their 'clients' (the farmers and other rural producers). Ministers have done likewise. The 'clients' would expect nothing less from the departments than the representation of 'our' interests within governments (Warhurst, 1990). Departments of agriculture have continued to 'represent' their clients, although there is now greater recognition that their clients are all taxpayers rather than farmers alone.

The fact that governments are responsible not only to urban dwellers but also to farmers and other rural residents and that those neutral, anonymous and apolitical public servants of departments of agriculture fiercely represent and defend the interests of their rural clients could be a most important factor responsible for today's very advanced Australian agriculture. The brief highlights of the evolution of Australian agricultural development in Chapter 2 showed that there were several 'small crises' in the history of Australian agriculture during each of which a relatively large number of small but financially vulnerable farmers emerged. Yet, every time these 'poor' farmers were helped either to overcome their financial difficulties to stay in farming or to quit farming with financial and other support, thanks to the governments that were responsible to these farmers and to the public servants of departments of agriculture who helped the farmers.

Had the Australian government been 'responsible' for urban residents only, and had public servants not been apolitical and unable to defend farmers' interests, what would have happened to those poor farmers? Many small, poor and inefficient farmers might not have been assisted to exit the land, but would have been left trapped and suffering. The existence of too many small farms would not have allowed Australian agriculture to achieve today's large-scale farming with its enviable efficiency.

3.4 Transparent Policy Process

In Australia, from government departments to committees and to small farmer organizations or groups, when policies need to be formed or reviewed, the process is generally transparent, inclusive and consultative. This ensures all those affected are given the opportunity to have a say in the formation or review of policies.

Transparency of the policy process is one of the most important institutional arrangements. In Australia, where practical, policy transparency is required by law, through federal and state 'freedom of information' legislations. One approach which Australia has practised to ensure that the policy process is transparent, inclusive and consultative is the 'public inquiry process' undertaken by the Productivity Commission (PC). The public inquiry process, most innovative and unique to Australia, is described briefly below.

The PC is the Australian Government's independent research and advisory body on a range of economic, social and environmental issues affecting the welfare of Australians. Its role is to help governments make better policies in the long-term interest of the Australian community. The philosophy lying behind the PC inquiry processes includes (Mauldon, 2009):

- **Transparency:** the processes adopted during an inquiry are fully open to the public. All submissions are public except those parts considered to be sensitive to the competitive performance of individual business activities. A public transcript is kept of public hearing proceedings.
- **Independence:** although the PC is required to have regard to any matters conveyed publicly by its minister, it is completely independent of any government direction about what matters it addresses or any conclusions it draws in fulfilling its policy guidelines.
- **Impartiality:** in fulfilling its policy guidelines, the PC is required not to favour any industry or consumer sector of the economy. Its sole requirement is to foster general efficiency of resource use among industries and firms within them and to maximize community well-being while recognizing that assistance may be needed by those finding it difficult to make needed adjustments to change.
- **Consistency:** although industries differ in their production and trading characteristics, the PC applies the same principles to all industries in its consideration of policy issues. This is true not only between industries, but also over time.

Public inquiries can vary greatly in issue and scope. The wide-ranging charter of the PC is reflected in the inquiry programme set by the Australian Government. The Treasurer, in consultation with other ministers (and state and territory governments where relevant), determines which inquiries the PC will conduct and the terms of reference for each inquiry.

An inquiry is initiated when a 'reference' is sent by the Treasurer to the Commission. The terms of reference describe what the inquiry will cover, the reporting date and whether public inquiries will be held. Most inquiries are specified for 9 or 12 months' duration. Some, however, may have a shorter term: for example, 3 or 6 months. References may cover any sector of the economy, address a particular industry or cut across industry boundaries, and involve wider social or environmental issues.

Public inquiries give the opportunity for all points of view in the community to be heard and considered. All individuals, firms, groups and organizations with an interest in an inquiry can participate in the inquiry. This may be through written submissions and attendance at hearings, workshops and other forums when these are held for the inquiry.

The success of each inquiry depends largely on the participation of people and organizations in the community. The Commission draws on the information it receives from participants, and information, research and expertise assembled from other sources to formulate the report and any policy recommendations. The views of interested parties are considered within an economic framework based heavily on a market economy where the role of government is aligned strongly with remedying market failure and where impacts are measured using standard welfare analysis.

The PC may make recommendations on any matters it considers relevant to the inquiry, but it must take into account its policy guidelines. In framing its recommendations, the PC is required to consider the economic, social, regional and environmental interests of the community as a whole, in addition to the interests of those most immediately and directly affected by the recommendations. The PC usually makes a draft report publicly available for scrutiny and comment before its final report is completed. This approach is designed to ensure consistent and open consideration of the issues. Once complete, the final report is forwarded to the government and awaits release by Parliament.

Final reports must be tabled in Parliament within 25 sitting days of the Treasurer receiving the report. Commonwealth, state and territory governments make the final decision on acceptance and implementation of the Commission's recommendations. Governments are not obliged to take the PC's advice. However, even when the PC's recommendations are not adopted, government policy making is often served better by the information gathering, public participation and scrutiny that the inquiry process stimulates. Also, governments have the responsibility to explain why the Commission's recommendations are not accepted.

Australians who care about public policy, including public servants, researchers and senior government officials, regard this public inquiry process highly. It ensures Australia's major policies are formulated to reflect best the interests of the broad community. The PC and its predecessors, the Industries Assistance Commission (IAC; January 1974–March 1990) and the Industry Commission

(IC; March 1990–April 1998), have conducted numerous inquiries. Appendix D, Inquiry Reports, in *From Industry Assistance to Productivity: 30 Years of 'The Commission'* (PC, 2003), provides a complete list of all the inquiries by the PC and its two predecessors until 2002. All inquiries since 2003 can be found at the website of the PC.

A large number of inquiries have been held concerning agricultural industries, especially during the 1980s, when agricultural deregulation gathered momentum. Such inquiries provided important guidance in Australia's effort to deregulate many aspects of its agriculture (see Chapter 5 for details).

As noted earlier, not all of the PC's recommendations are necessarily accepted by the governments. The governments of the day may choose not to accept certain recommendations for various other considerations. None the less, the governments do ultimately accept many recommendations, though sometimes after some delay.

Being transparent is a laudable ambition of policy processes at various levels in Australian society. Transparency of how important decisions are made is expected for all levels of government departments, committees and even small farm groups. This helps decision makers, no matter whether they are senior government officials or leaders of a small farmer group, to reduce or avoid misallocation of resources, to be fair, to be socially responsible and to avoid abusing their power.

When any inquiries concerning agricultural industries are held, farmers, rural industry leaders and any other persons who are interested in the matter have the opportunity to voice their concerns, opinions or suggestions. As such, the rural community is not just passively given any policies in which they have no inputs but is actively involved in the making of policies that affect their farming operations and their lives.

3.5 Agricultural Services

Agricultural service institutions have been well developed and have played an important role in the development of Australian agriculture. In Australia, agricultural services are provided by both the public and private sectors. The private agricultural service industries have been well established. Today, most agricultural services are still provided by public sectors, while the role of private sectors in the provision of agricultural services has been increasing.

3.5.1 Public agricultural services

The provision of services to agriculture is shared between the Commonwealth and state/territory governments. Broadly, services provided for agriculture include regulatory services (such as quarantine, disease and pest control, and biosecurity) and agricultural research, development, extension and, to some extent, education. The Commonwealth government provides such services that are of broad national relevance. It also plays a coordinating role in areas involving interstate cooperation. State governments provide such services within their states. All state departments of agriculture also conduct research that is of regional application and importance.

Control over the movement of plants and animals has always been an important aspect of regulatory administration in Australian agriculture at both the federal and state levels. Australia is free of many serious diseases and pests of plants and livestock. Imports of many types of livestock are prohibited. There are also very stringent regulations governing the conditions under which seed, plants and plant products can be introduced into Australia. The Australian Quarantine and Inspection Service, which is part of the Australian Government Department of Agriculture, Fisheries and Forestry (DAFF), is responsible for quarantine policy in respect of entry of livestock and plants from overseas. At the state level, some local restrictions, administered by state governments, are also in place on the movement of plants and animals between states.

Prior to the 1980s, government services were directed primarily towards the improvement of farm production, production efficiency and market access. While this continues to be important, there have been, since the 1980s, increasing emphases on environment protection, sustainable farming and biosecurity.

3.5.2 Private agricultural services

In Australia, the private sector is involved in the provision of many services, except most regulatory services, to the agricultural industries. Major such services include:

- the supply of agricultural inputs, including seed, fertilizers, chemicals (such as insecticides, herbicides, fungicides);
- animal feeds;
- agricultural machinery;
- harvesting services;
- processing and marketing;
- financial services;
- production and management consultancy services;
- transportation services; and
- agricultural insurance services.

Basically, all the services that farmers need pre-, during and post-production are provided by the private sector. The private service network covers the whole country and such services are just a telephone call away from farmers.

The significant involvement of the private sector in providing consultancy services in Australia is attributable partly to the privatization of many extension services that used to be provided by state governments. State departments of agriculture nowadays provide limited extension services and many such services have been shifted to private enterprises. The private sector has also become engaged increasingly in agricultural R&D services (Keogh, 2011).

3.6 Other Important Institutions

There are other institutional arrangements that facilitate agricultural development. These arrangements are based on the principle that citizens living in rural areas are not discriminated against. As such, farmers are free to form political parties or other organizations to represent their interests. They enjoy the same social security protection and medical care service as citizens in urban areas. Essential infrastructure and telecommunications services are provided for rural residents no matter where they live. Farmers are also free to choose where they want to live and work: live and work on a farm, live in a town and work on a farm, or quit farming and move to an urban area to work and live. Such non-discriminatory institutional arrangements help those who wish to exit from farming to leave the land, enabling remaining farmers to expand and benefit from economies of scale.

3.7 Concluding Comments

As in other countries, institutional arrangements have contributed to the shaping of Australian agriculture. Not all such institutions are, or can be practically, addressed in this chapter. Some are discussed in the following chapters, such as institutional arrangements for agricultural R&D and for fostering farmer entrepreneurship. Addressed in this chapter are some of the institutional arrangements that are fundamental (the constitution, land tenure and responsible government), general (agricultural service institutions, equal treatment of rural residents) or unique to Australia (the public inquiry process by the PC). These five aspects of institutional arrangements, together with various others, have played a major role in determining the shape of today's Australian agriculture.

The constitutional division of powers between Australia's federal and state governments (Section 51) and the Constitution's insistence on freedom of interstate trade (Section 92) have had a substantial influence on the development of agriculture and agricultural policy. Due to these two constitutional clauses, at times it was difficult and time consuming for the federal government to implement policies that required a national approach. On other occasions there were problems concerning interstate trade. However, there is little evidence to suggest that Sections 51 or 92 have had serious adverse consequences for Australian agriculture. Instead, the institutional arrangements resulting from these two clauses give states autonomy to administer their agricultural sectors in a way that utilizes best the comparative advantages of their states to develop their agricultural industries. Different approaches in different states tackling similar issues also enable states to learn from each other.

The high proportion of public land ownership clearly did not have much negative impact on Australia's agricultural advancement. Australia's experience shows that so long as farmers' rights to farm the land are protected by law, whether the land is owned publicly or privately does not seem to matter much at all.

In Australia, governments are responsible to the parliament and to the people who elected the parliament. Universal franchise means that governments in Australia are responsible not only to urban people but also to rural people. Transparency of the policy process has become a common feature of Australian policy process at various levels. Australia's practice of having a public policy inquiry is a great innovation and success. Public inquiries enable farmers' points of view to be heard and considered.

Both the public and private sectors contribute to a well-established agricultural service institution in Australia. Over the recent past decades, the public sector has reduced its provision of extension services. Services provided by the private sector have increased. The private sector will continue to provide many essential services to rural industries. Such a well-established agricultural service sector contributes to the efficiency of Australian agriculture.

In terms of getting institutions right to serve the agricultural industries, Australia's experiences may be summarized as follows:

- Non-centralized administration of agricultural development allows states to utilize best the comparative advantages available to their states.
- Land ownership does not have to be a deterrent to agricultural development if the lessees' rights are protected adequately by law.
- Governments should be responsible to both urban and rural residents, and rural residents should be treated equally.
- The policy process should be transparent.
- A well-established agricultural service sector is necessary to support agricultural advancement.

These experiences can be of great relevance to many other countries, particularly those developing countries where institutions for agricultural development are absent or weak. Admittedly, different countries have different institutions and different ways of doing things. None the less, looking into Australia's experience in setting its institutions for agricultural development may provide valuable revelations to help other countries to make changes.

First, countries with diverse agricultural resources in different regions may consider changing centralized administration to give autonomy to state or provincial governments to administer their own agricultural development. This can help them to utilize their comparative advantages better. In India, state governments are given autonomy to administer their own agricultural development, enabling each state to capitalize on the comparative agricultural advantages endowed to them (V. Gandhi, Indian Institute of Management, 11 May 2012, personal communication). In China, on the other hand, the central government's Ministry of Agriculture supervises the activities of provincial departments of agriculture closely. This highly centralized approach reduces the flexibility for different provinces to deal with or focus on issues peculiar to their own provinces. Often, the comparative advantages of their provinces cannot be brought to play.

Second, farmers' rights to use land, either owned or leased, have to be protected adequately by law. Otherwise, no matter whether the land is owned privately or publicly, agricultural development will suffer.

Third, farmers must be treated equally in society. Treating farmers as second-class citizens is not only morally unacceptable but also hinders a country's agricultural development significantly. When agricultural productivity improves, rural people should be assisted to leave the land rather than holding them institutionally on the land. Confining farmers on land is discriminatory but, more importantly, prevents less efficient farmers from exiting, depriving more efficient farmers of opportunities to expand. This reduces the opportunity to allow rural adjustments to take place continuously and gradually. Rather, needed adjustments are prevented, postponed and accumulated. The accumulated adjustment needs may become so enormous as to be difficult to tackle. When too many people are kept in rural areas, the only possibility is that farmers become poor. A rural population with limited purchasing power to buy the goods and services produced in the urban system limits urban employment expansion. In this regard, it is the government's responsibility to have policies to treat rural people equally with the urban population. Policies should be devised to encourage farmers to leave the land and, where appropriate, necessary support should be provided to those farmers who want to leave the land but are less able to due to financial or other difficulties.

Fourth, policy processes need to be transparent and involve farmers and others in their development. The Australian PC experience to ensure policy process is transparent, inclusive and consultative is of great value and relevance as a model to other countries, both developing and developed. Being transparent in policy development helps keep a government honest and responsible to its people. It is the view of this author that every country should establish an organization similar to Australia's PC. It is interesting to note that, in 2011, New Zealand established a PC similar to Australia's. The New Zealand PC is an independent Crown entity that began operating on 1 April 2011 (NZPC, 2011).

Last, developing agricultural services should attract more attention. For many developing countries, the provision of agricultural services is generally weak in both the public and private sectors. In view of the importance of agricultural services to agricultural development, more efforts will be needed to establish and fine-tune their agricultural service sector so that services are provided competitively and with appropriate legal protection for buyers and sellers making market transactions.

Note

[1]Selling leased land is quite complex. It depends on the type of lease and the period for which the lease is granted. If the permits to occupy are short term, for example less than 2 years, then the lease cannot be transferred from the leaseholder's name. Other leases such as grazing homestead pastoral leases can have a period of up to 99 years, so purchasers pay a price to take the lease over. However, the purchasers never fully own the land but continue only to lease the land until the lease period runs out. In the case of long-term leasehold, for example a 99-year lease, then the leasehold land is sold virtually no differently from freehold land.

References

Campbell, K.O. (1980) *Australian Agriculture: Reconciling Change and Tradition*. Longman Cheshire, Melbourne, Australia.

Cooray, M. (1995) Property Ownership in Australia (http://www.ourcivilisation.com/cooray/rights/chap813.htm, accessed 16 October 2009).

DERM (2011) Leases (http://www.derm.qld.gov.au/land/state/leases.html, accessed 10 November 2011).

DIA [Department of Indigenous Affairs] (2010) Land Facts. Western Australian Government (http://www.dia.wa.gov.au/en/Land/Land-Facts/, accessed 2 September 2010).

Jessup, J.E. and Dun, R.B. (1982) Organization and administration. In: Williams, D.B. (ed.) *Agriculture in the Australian Economy*, 2nd edn. Sydney University Press, Sydney, Australia, pp. 106–124.

Keogh, M. (2011) Private sector investment in agricultural research and development in Australia. *AFBM Journal* 8(2), 14–19.

Lewis, J.N. (1967) Agricultural price policies. In: Williams, D.B. (ed.) *Agriculture in the Australian Economy*. Sydney University Press, Sydney, Australia, pp. 299–314.

Lloyd, A.G. (1982) Agricultural price policy. In: Williams, D.B. (ed.) *Agriculture in the Australian Economy*, 2nd edn. Sydney University Press, Sydney, Australia, pp. 353–382.

Mauldon, R. (2009) The Industry Commission's 1991–92 inquiry into the Australian sugar industry. Presentation to the APEC training programme in resolving policy conflict: *Food Security, Structural Reforms and Food Price Inflation*, 17–24 June 2009, Melbourne, Australia.

NZPC [The New Zealand Productivity Commission] (2011) The Commission. NZ Government, Wellington (http://www.productivity.govt.nz/about-us/the-commission, accessed 8 March 2012).

PC [Productivity Commission] (2003) *From Industry Assistance to Productivity: 30 Years of 'The Commission'*. Australian Government, Canberra.

PC (2010) The Public Inquiry Process (http://www.pc.gov.au/about-us/inquiryprocess, accessed 27 July 2011).

SCoPI [Standing Council on Primary Industries] (2012) Background to the Councils (http://www.mincos.gov.au/background#top, accessed 6 April 2012).

Warhurst, J. (1990) The politics of rural Australia: rural industries and rural areas. In: Williams, D.B. (ed.) *Agriculture in the Australian Economy*, 3rd edn. Sydney University Press, Sydney, Australia, pp. 109–126.

4 Protecting Farmers' Interests

In the previous chapter, it is noted that it is important to protect farmers' interests. After all, all farm production is carried out by farmers. If they are not protected and encouraged to work the land hard and smart, agricultural development is bound to suffer. This chapter demonstrates how farmer interests in Australia are protected.

There are several mechanisms with which farmers' interests are protected in Australia. These include rural politics, farmers' own organizations and overall conducive institutional arrangements and government policies.

4.1 Rural Politics

Rural politics is one of the key factors that have contributed to the effective protection of farmers' interests in Australia. Rural politics is concerned with ensuring that matters affecting rural industries and the welfare of rural residents are channelled through and properly addressed. The importance of rural politics lies in that: (i) rural concerns are voiced adequately in society; (ii) rural interests are represented at various levels, for example, state and federal; and (iii) rural concerns and interests are translated into policy inputs, which are taken into account in government policy making. Hence, rural politics helps agricultural and rural development to gain deserved attention from society and helps a country to avoid neglecting the needs of rural development. As such, rural politics plays an important role in assisting a country to achieve a more balanced societal and economic development.

The starting point in understanding rural politics has to be general political institutions and processes, much of which has been dealt with in Chapter 3. In this chapter, we focus on how people in rural Australia become involved in politics (Section 4.1.1) and how rural politics has generated impacts that have helped to improve the welfare of rural Australians in general and to advance Australia's agriculture in particular (Section 4.1.2).

4.1.1 Rural politics in action

Rural Australians become involved in political interactions with state and federal governments, both as individuals and collectively. As individual citizens, country people participate in the election of state and federal parliaments. Collectively, through political parties and pressure groups, they apply organized pressure in pursuit of their own ends.

Individuals in political interactions

Rural attitudes and political behaviour are generally different from urban ones. People living in rural areas and country towns hold beliefs that are more conservative than those held by their city counterparts, no matter whether they are farmers or factory workers. Their conservative beliefs usually lead them to supporting conservative political parties, that is the Liberal Party or the National Party (previously the Country Party). Generally, the Liberal and National Parties fare better in country towns and, especially, in rural areas. The Australian Labor Party (ALP), on the other hand, performs poorly in rural areas other than country towns. However, this does not necessarily mean that the ALP cannot win in rural electorates. In fact, successful ALP governments have always won some country electorates. In short, individuals in rural areas participate in politics through electing who is going to represent them in local, state and federal governments.

Political parties and pressure groups

In rural areas, the two major vehicles for collective political action have been farmer organizations and the National Party. Farmer organizations represent a major force for collective political actions and have effectively lobbied, advocated and led many issues concerning farming needs and interests. There are a large number of farmer organizations in Australia at various levels (from the grass-roots level, to state and national levels) and of different types (e.g. cooperatives, associations, councils and federations) for all kinds of industries and farm commodities. More details about farmer organizations in Australia are given in Section 4.2.

The other major force for collective political action has been the National Party, previously the Country Party. Members of the National Party in Australian parliaments have represented farmers' interests and brought issues concerning the farming community to parliamentary debates. The National Party has also been in various governments at both the state and federal levels, with its members assuming various senior government positions. This allows the National Party to have more input into the formation of policies that affect the farming and rural community.

Very close connections existed between farmer organizations and the Country Party. Farmers' and graziers' associations created the Country Party in 1920. As a result, the Country Party was often seen as a farmers' party in Australia. The party was established as a major contestant in state elections from 1920 and

Commonwealth elections from 1922. Various farmer organizations used to be affiliated to the party. By the late 1940s, all farmer organizations had broken official links with the Country Party (Connors, 1996, p. 46). The divorce between farmer organizations and the Country Party was one of mutual consent. For the Country Party to have a future, it had to stop concentrating on the farm sector and move from being a sectional party to a regional party. On the other hand, the split helped farmer organizations to improve their communications and dealings with all governments, especially the Labor government (Connors, 1996, p. 47).

Despite the non-official affiliation between the two, they still work closely together. Often, farmer organization members are appointed to various key positions that deal with rural issues when the National Party is in coalition with the Liberal Party. Leaders of the National Party also emphasize the closeness of the relationship between the party and farmer organizations and their reliance on advice from farmer representatives in domestic and international negotiations.

4.1.2 Influence of rural politics

Farmers in Australia have been well involved in Australia's politics and, in turn, their interests have been well represented in the process of government policy making. To see the influence of rural politics on government policies, and subsequently farmers' economic and social welfare, we could examine the results that each of the three major players in rural politics can and have achieved: namely, farmers, the National Party and farmer organizations.

Individual farmers

Farmers in Australia choose through elections who represents them politically in governments, although there are now few seats where the farm population is larger than the non-farm population. If such an elected representative was perceived not to work to the expectations of the electors, farmers would vote this person out and choose someone else to represent their interests. Sometimes, individual farmers themselves get elected and then defend farmers' interests at various levels of government. Farmers can also form their own organizations to represent to government their interests in production and marketing (as shown in Section 4.2). In addition, elected farmer representatives from various industries can be appointed to government bodies such as statutory authorities, who can make representations directly during government legislation processes.

The National Party

Warhurst (1990) suggests that there are three ways to measure the influence of the National Party in its quest to represent farmers: (i) the number of members of the National Party in parliament; (ii) the National Party in government; and (iii) the impact of National Party policies. According to Warhurst: (i) there have always been plenty of National Party members in Australian parliaments who are themselves primary producers; (ii) the National Party has also been in various governments at both the state and federal levels, either in coalition (at both the state and federal levels) or by itself as a major party at the state level; and (iii) the

National Party did establish a system of support for rural industry. It was influential at key times in determining policy within the coalition on matters that were of particular concern to agricultural industries. In addition, MPs of the National Party have been appointed Deputy Prime Minister or Deputy Premier/Premier or senior ministers in the cabinets when the Coalition is in power. Hence, the National Party is indeed a strong representation of farmers.

Farmer organizations

Since the 1890s when the first farmer organization was established, Australian farmers have formed many kinds of their own organizations at different levels for various industries and commodities. As rural pressure groups, their achievements in getting representation in parliaments, forming a political party and sharing the government benches are 'notable', according to Connors (1996, p. 44). Addressing in detail the influence that many farmer organizations have had on issues that affect farmer interest is not practical, but some examples are given. For instance, the influence of the National Farmers' Federation (NFF) on protecting Australian farmers, as well as its positive impact on the nation's economic, environmental and social well-being, is given (see Section 4.2.2).

4.2 Farmer Organizations

The first farmer organizations in Australia were those formed by graziers towards the end of the 19th century.[1] They had their origin in the industrial unrest associated with the depression of the 1890s. During this period, the Amalgamated Shearers' Union (ASU) and the Queensland Shearers' Union practically demanded that only unionist shearers should be employed by Queensland pastoralists, and in May 1890, the Australian Labour Federation declared the intention of the waterside unions to block the shipment of wool shorn by non-unionists.

Unorganized pastoralists were unable to resist the union demands and some agreed that only union shearers would be employed. Alarmed by this decision, groups of pastoralists in Queensland and in the 'southern colonies' saw the need for statewide bodies and a federal council to protect their interests.

The pastoralists quickly established their own (employers') unions in NSW, VIC and SA. In December 1890, the United Pastoralists' Association of Queensland was formed. At an 'Intercolonial Conference' held in Sydney in December 1890, the Pastoralists' Federal Council of Australia was established. As a result, the influence and sphere of operations of the combination extended from the Gulf of Carpentaria to Bass Strait, from the seaboard of QLD and NSW to the western boundary of SA.

Meanwhile, in July 1890, a Shearers' Manifesto had been issued by the ASU appealing to every unionist 'to draw such a cordon of Unionism around the Australian Continent as will effectually prevent a bale of wool leaving unless shorn by Union Shearers'.

The pastoralists were concerned with the 'freedom of contract', that is, the right to employ whom they wished. In January 1891, shearers were asked to sign on under an agreement form prepared by the Pastoralists' Federal Council, which allowed non-unionists to work in sheds alongside unionists. Many shearers refused to sign on. A great struggle ensued in QLD and NSW: pastoralists tried to continue shearing without union labour and members of the ASU attempted to disrupt operations.

This culminated in a conference in Sydney in August 1891 between the Pastoralists' Federal Council and the ASU at which it was ceded 'that employers shall be free to employ and shearers shall be free to accept employment, whether belonging to shearers' or other unions or not, without favour, molestation or intimidation on either side'. Later, an agreement for the conduct of shearing was made and subsequently it was adopted universally throughout the colonies. In these early years, the industrial and political structure of employer and employee organizations found their beginnings. In later years, it was problems in marketing and differences in the interests of various groups of producers that led to the formation of new producer organizations.

4.2.1 Current status of farmer organizations

After over 100 years of experience in farmer organization, Australian farmers today are now well organized. All states and almost all agricultural industries have farmer organizations and, in some cases, individual agricultural products have their own organizations. Each state has relevant acts to guide the formation and operations of farmer organizations. The governments do not intervene in any way in the operations of farmer organizations and they provide no financial support either. Farmer organizations have to survive on the fees contributed by members. However, if they are able, farmer organizations are allowed to invest to earn a profit to support their activities and they are also allowed to transform into other structures such as companies or corporations when they become strong enough.

Types

Several types of farmer organizations exist in Australia today. Typically, these include cooperatives, associations, councils and federations.

COOPERATIVES A group of producers work together for a common interest. In most cases, such cooperatives are product specific and are formed to market the products of members: see for example, the Molong, Manildra and Cumnock (MMC) Co-op in Section 4.2.2. Most of them are confined within a local area. Interstate or national farmer cooperatives are not commonly seen in Australia. Indeed, cooperatives have not been used extensively by Australian farmers. Campbell found it 'difficult to pinpoint the reasons'. He speculated that:

> It may be that the long-standing Country Party view that farmers are as qualified to market their product as they are to produce it has restricted the success of cooperatives. A still further explanation may lie in the fact that marketing boards, which might, not unjustly but rather incongruously, be described as 'compulsory cooperatives' have served the function of cooperatives at least on the marketing side.
>
> (Campbell, 1980, p. 110)

When Zhou (2004) was examining China's agricultural cooperatives, he argued that the need for such cooperatives was related to the provision of agricultural services. Using Australia as an example, he pointed out that Australia's agricultural services were well developed. As a result, many functions performed by cooperatives might have been replaced by such services, hence reducing the need for cooperatives.

ASSOCIATIONS Associations generally look after broader issues that concern the interests of agricultural producers other than a single product, for example the NSW Farmers' Association. Some organizations, however, do use the term 'association' when they are concerned with only one product, for example the Northern Territory Cattlemen's Association and the Ricegrowers' Association of Australia (RGA).

COUNCILS This term is often used when a farmer organization is concerned largely with only one product. They are generally at the national level. Examples include the Sheepmeat Council of Australia, the Goat Industry Council of Australia and the Cattle Council of Australia.

FEDERATIONS Federations are also concerned with broader issues that affect the interests of agricultural producers. Generally, they cover much larger areas, for example a state or the whole nation. An example of the former is the Victorian Farmers' Federation. The NFF is an example of the latter.

Levels

Farmer organizations in Australia may be viewed from three levels: the grass-roots level, the state level and the national level.

GRASS-ROOTS LEVEL At the grass-roots level are those small-scale and localized organizations that deal with specific matters that are of interest to members. See, for example, the MMC Co-op in Section 4.2.2.

STATE LEVEL There are farmer organizations that operate at the state level concerning single industries/products, such as the Murray Goulburn Co-operative (dairy). Each state also has a state-level body representing all farmers in their state for broader issues of interest. These bodies are:

- AgForce Queensland.
- Northern Territory Cattlemen's Association (NTCA).
- NSW Farmers' Association.
- South Australian Farmers' Federation (SAFF).
- Tasmanian Farmers and Graziers Association (TFGA).
- Victorian Farmers' Federation.
- Western Australian Farmers Federation.

NATIONAL LEVEL At the national level, there are a number of farmer organizations representing members nationwide on an industry basis or on a single commodity basis: for example, the Cattle Council of Australia and Wool Producers Australia. The most important farmer organization at the national level is the NFF.

Numbers

Attempts were made to find the total number of farmer organizations in Australia and to sort them according to organization types, different industries and different states and territories. Surprisingly, such statistics are not readily available. Help was provided by the NFF, whose staff, after inquiries, came to the conclusion that such data were not readily available. It was explained that part of the reason could be the diversity and complexity of interests for which farmer organizations try to cater, making it difficult to collect such statistics (S. Nelson, NFF, 28 October 2011, personal communication).

4.2.2 How they operate

Four farmer organizations, at the grass-roots level, industry specific, at the state level and at the national level, are used to highlight how farmer organizations operate in Australia. They are the MMC Co-op, the RGA of Australia, the NSW Farmers' Association and the NFF.

Molong, Manildra and Cumnock Co-op

Prior to 1989, wheat marketing in Australia was monopolized completely by the Australian Wheat Board (AWB). Domestic wheat marketing was liberalized in 1989. To cope with the competitive market, eight wheat growers from the Molong, Manildra and Cumnock region (of the central west of NSW) got together and proposed to form a marketing cooperative to help individual wheat growers in their region to achieve a better price for their wheat. The proposal was well received. In 1991, the MMC Co-op began operating (Whiteley and Zhou, 1998).

The MMC Co-op each year decided a total amount that would be marketed for its members. Members decided how much wheat they planned to sell through the co-op to make up this total amount. Before wheat harvests, the total amount was announced for tendering from buyers. The co-op guaranteed the successful tender the supply of a minimal amount. In the meantime, the successful tender was also required to agree to buy a reasonable amount of extra production if the collected total was slightly higher than the planned total. This simple method worked well for the co-op.

The operation of the co-op was extremely flexible, simple and low cost. To become a member of the co-op, one needed to pay only AUS$35 membership fee per annum. Joining was completely voluntary. One could join or exit at any time. Each member decided how much of their harvest was to be sold through the co-op and the co-op reserved the right to accept or refuse the amount proposed by a member.

The MMC Co-op had a small board of directors, composed of the members. None of them was compensated for their time and efforts. What they did was purely voluntary. Because it was small, there was no need to have full-time staff for operations. Hence, its operation costs were minimal. Membership contributions were sufficient to cover all the operation costs.

The MMC Co-op stopped operating in 2008. Part of the reason was the severe droughts in most of the years during the first decade of the 2000s. The harvests were generally poor and wheat growers were generally able to sell for a good price. The other reason was that over the years since the deregulation of the domestic wheat market in the late 1980s, wheat producers gradually established silos on their farms and they could hold on to their harvests to wait for a good price (W. Whiteley, Molong, NSW, 16 December 2009, personal communication).

Ricegrowers' Association of Australia

The RGA represents farmers in the main rice growing areas of NSW and VIC, with headquarters in Leeton, NSW.[2] It was formed in 1930 to unite the small group of pioneer rice growers into an effective and cohesive force. It represents over 1500 voluntary members and is the collective voice of Australian rice growers.

Rice growers in Australia today perhaps need the association more than ever to represent their interests because of the water scarcity facing Australia and the community's perception that rice production is most water thirsty. The RGA has been working hard to diffuse water usage concerns from the community. It leads growers on issues affecting the viability of their business and communities. It plays an important role in: (i) developing and implementing policy for the rice industry that is in growers' best interests; (ii) representing the interests of rice growers to federal, state and local governments and their various agencies, the NFF, SunRice, other interest groups and to the community generally; and (iii) servicing the specific needs of individual members.

The RGA has eight branches (Yanco, Mirrool, Coleambally, Berriquin, Deniliquin, Wakool, Hay and Victoria). Branches meet at least twice each year. Members elect their Branch Executive and delegates to the Central Executive. An annual conference of members is held each year to discuss issues brought forward from branches and the Central Executive.

The Central Executive is the peak decision-making body of the RGA. It considers matters brought forward by members through their branches, and other matters of importance. There are six committees under the Central Executive, each having a particular focus. The six committees are the Water Committee, the Farm Business Committee (incorporating the Climate Change Taskforce), the Industry Affairs Committee, the Sustainability Committee (incorporating the Environmental Working Group), the Finance and Organization Committee and the Crop Protection Committee (incorporating the Biosecurity Taskforce). The association is serviced by a small secretariat that includes the Executive Director, the Policy Officer, ECP Regional Coordinators and the Office Manager based in Leeton, NSW. The organizational structure of the RGA in 2012 is shown in Fig. 4.1.

The RGA has a Strategic Plan that assists the Central Executive to focus efforts and resources in three keys areas – core, internal and external business. Core business strategies and actions are designed to provide:

- grower support and services;
- representation on a range of water issues dealing with access, sharing and responsibilities;

- single desk rice marketing arrangements;
- target R&D programmes;
- sustainable environments through the RGA's Environmental Flagship programmes of Biodiversity Strategy and Plan, Greenhouse Challenge and Environmental Champions Program;
- assist members to achieve better business management programmes and outcomes;
- contribute to social goals to ensure healthy and vibrant communities; and
- promote the industry regionally and across Australia.

Internal business activities focus on internal communication and organizational capability and relate mainly to the functions of the Central Executive and staff. External business activities focus on strategies and actions designed for communication with members, public perceptions, understanding the rice industry and influencing target audiences such as key decision makers.

Individuals who are producers of rice, their family members or share-farmers and other persons approved by the Association Executive, are eligible for membership of an Association Branch. To become a member of the association, an annual contribution is required. The contribution rate is set annually by the RGA annual conference. The current single membership fee (2011–2012) is AUS$360 (GST inclusive). If more than one from the same family wishes to become a member, the extra payment is AUS$66/person. A substantially reduced fee (AUS$66) is set for retired and young rice growers to encourage their participation. For associate members, the annual contribution is AUS$164.

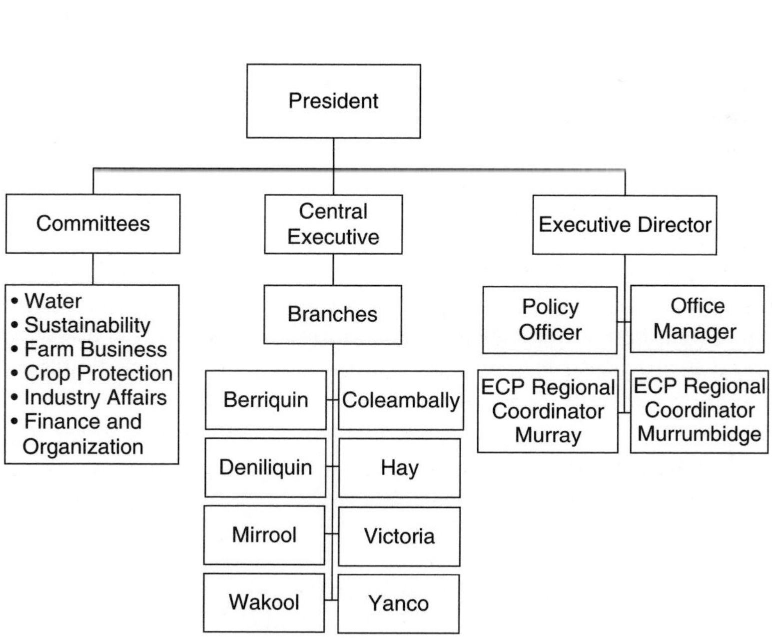

Fig. 4.1. Organization structure of the Ricegrowers' Association of Australia (based on RGA, 2012).

NSW Farmers' Association

The NSW Farmers' Association can trace its history back to the 1890s with the formation of the Pastoralists Union and the Farmers and Settlers Association.[3] These bodies later became the Graziers Association and the United Farmers and Woolgrowers Association. In 1978, these bodies and the Graziers Association of the Riverina merged and from 1985 adopted the name of the NSW Farmers' Association. It has since been the representative voice of farmers in NSW. It promotes agriculture and the best interests of farmers and rural communities across NSW.

The NSW Farmers' Association is regarded as one of the most successful state-level associations in Australia. Its success is based on its long-standing policy of operating as a 'grass-roots' organization by maintaining the closest possible communication links between all members. The other major reason for its success is the innovative style of association management. The NSW Farmers' Association forms partnerships with those who directly serve rural needs. In this way, the association is in a better financial position, thus, offering members lower association fee contributions. This encourages more farmers to be part of the association, making it even stronger financially. In the meantime, members can obtain services from the association's partners at a discounted price, benefiting both the members and the association partners. Another example of benefits to members is the establishment of the Country Law Firm Member Benefit Program. This is a specific member benefit in the area of legal services. Members have access to regionally accessible legal advice, at competitive rates, from firms with a commitment to rural and farming issues. Most of the firms participating in the programme are Law Society best practice firms or Law Society accredited specialists. The programme is an additional, separate and complementary programme to the limited free Internet legal advice currently available to members.

The association welcomes businesses and individuals with an interest in farming or rural living. In addition to the full producer membership, it also has devised several other categories to encourage wider participation at a lower or zero cost. For full producer membership, the fee is charged according to farm annual turnover. This membership applies to the owner or manager of a farm enterprise. Full producer members are also encouraged to pay an additional industrial relations (IR) fee, which is again varied according to annual farm turnover. When needed, an IR Essentials package will be provided to a full producer member, containing IR information and advice. It is a type of insurance and it is provided because each year more and more rural employers face legal action over IR and OH&S (occupational health and safety). The fees for various membership categories and what services a member is entitled to are given in Table 4.1.

National Farmers' Federation

Before the NFF came into existence, there were deep divisions between farmer organizations. When Tom Connors was an agricultural writer in the 1960s, he was often both intrigued and confused by such divisions (Connors, 1996, p. 1). There was the need for a unified position with 'cohesion of purpose' to achieve

Table 4.1. Membership fees and member entitlements of the NSW Farmers' Association (NSW Farmers' Association, 2012).

Part A: Full Producer Member (PC)

Category	Membership fee (AUS$)	Industrial relations fee (AUS$)
Category 1: less than AUS$100,000	370	140
Category 2: AUS$100,001–500,000	520	155
Category 3: AUS$500,001–1,000,000	880	215
Category 4: AUS$1,000,001–3,000,000	1,330	435
Category 5: Over AUS$3,000,000	Contact Member Service Centre on 1300 794 000	

Part B: Other membership categories

Category	Membership fee (AUS$)
Second vote (SV): has a proprietorial or income interest in a Full Producer Member's business.	100
Additional member (AD): a second partner or company member where one partner is a Full Producer Member.	100
Associate member (AS): has an interest in agriculture.	100
Metro membership (MM): lives in a metropolitan area and is a supporter of agriculture in NSW.	100
Retired member (R): has an interest in agriculture.	100
Young farmer (YF): 18–35-year-old working in the agricultural industry.	0 (1st year only)
Student member (ST): a full-time tertiary student.	45

Part C: Membership inclusions

	PC	SV	AD	AS	MM	R	YF	ST
Voting rights	√	√	√					
Industrial relations (IR Basics)	√							
Member representation	√	√	√	√	√	√	√	√
Membership privileges	√	√	√	√[a]	√	√	√	√
Communication services	√	√	√	√	√	√	√	√
Support from MSC[a] and RSM[b]	√	√	√	√	√	√	√	√
Industry training discounts	√	√	√	√	√	√	√	√
Legal support services	√	√	√			√		
Tax advice	√	√	√			√		
Website unrestricted access	√	√	√			√	YF site	ST site

Notes: [a]MSC = Member Service Centre; [b]RSM = Regional Services Manager.

better real gains for farmers. This led to the formation of the NFF in July 1979. The informative book by Connors (1996), *To Speak with One Voice: The Quest by Australian Farmers for Federal Unity*, provides valuable historical accounts about the formation process of the NFF.

The NFF is a federation of independent, state and commodity organizations, funded voluntarily by farmers and representing all major pastoral and cropping industries in Australia. It brought primary producers together for the first time as a single, federal body. Since 1979, the NFF has helped Australian farmers to speak with one united voice.

The NFF consists of state farm organizations, national commodity councils and associate members. NFF members in 2012 are given in Table 4.2. Currently, SAFF is not a member of the NFF, the only state farmer organization that does not belong to the NFF. SAFF withdrew from the NFF in 2005. At that time, the membership fee for the SAFF was AUS$500,000, which was thought to be too high (the NFF has since had some changes in fee structures so the fees are now much less). The other major reason for withdrawal was policy differences. It was also believed that the NSW Farmers' Association had too much dominance, making the agenda of the NFF less relevant to SA (C. Vincent and D. Crabb, SAFF, 15 February 2011, personal communication).

The NFF only takes organizations as its members and does not have individual farmer members. Operating under a federated structure, individual farmers join their respective state farm organization or commodity council. By doing so, individual farmers contribute to and support the NFF, because part of the levies received by state farm organizations and commodity councils are used to fund the federation. State farm organizations and commodity councils collectively form the NFF.

Each of the state farm organizations deals with state-based 'grass-roots' issues and each of the national commodity councils deals with commodity-specific issues.

Table 4.2. Members of the National Farmers' Federation, 2011–2012 (NFF, 2011, pp. 56–57).

State farming organizations	Commodity councils	Associate members
AgForce Queensland	Australian Dairy Farmers' Limited	Australian Livestock Exporters' Council
Northern Territory Cattlemen's Association	Dried Fruits Australia	Australian Livestock and Property Agents Association
NSW Farmers' Association	CANEGROWERS	Australian Veterinary Association
Tasmanian Farmers and Graziers Association	Cattle Council of Australia	Beechworth Honey Pty Ltd
Victorian Farmers' Federation	Cotton Australia	Corporate Agricultural Group
Western Australian Farmers Federation	Ricegrowers' Association of Australia	Goat Industry Council of Australia
	Sheepmeat Council of Australia	GrainCorp
	Wool Producers Australia	Pastoralists' Association of West Darling
		Ridley Corporation

The NFF represents the agreed imperatives of all at the national level. It deals with issues that impact on a range of commodities, or over a number of states. Its aim is to create the most favourable economic environment in which rural industries may prosper and returns to farmers may be maximized.

The NFF's priorities are the broad national issues: economic policy, IR, trade, transport and the environment. It also takes a leading role in projecting the importance of agriculture to the national economy.

The NFF also represents farmers and rural and regional community interests in other areas such as social welfare, education and animal welfare. This involves contact with federal government ministers, the federal opposition, parliamentary and other committees of inquiry, and other business and community groups. The NFF has close working links with government departments – at state and federal level and semi-government bodies.

As well as promoting rural policy issues to decision makers in Canberra, the NFF recognizes the need for effective communication with the whole community. It publishes an annual report, outlining its activities, as well as a range of specific-interest publications, and distributes information on agriculture and issues of importance to the rural sector through the national and regional media. The NFF also regularly issues public statements on issues and events that have an impact on farmers and the communities in which they live.

The NFF's member organizations elect their representatives to the NFF Members' Council. The council is the supreme consultative forum for the NFF. The council elects the directors and office holders (President and Vice-President) from nominations received from members, and determines:

- the key priorities for the NFF's Strategic Plan;
- elects committee chairs;
- receives the budgets and financial reports;
- determines policy on issues or positions that are outside of the guidance previously agreed; and
- generally undertakes all functions and powers normally exercised by members or shareholders under the Corporations Act.

The Members' Council meets formally three times a year to discuss and formulate NFF policy. Several special meetings may also take place as needs arise.

There are committees and taskforces within the NFF. These internal committees and taskforces provide the Members' Council with high-level detail and advice on all policy issues. A chairperson is elected annually for each committee by the Members' Council in June. Currently, the standing committees charged with major portfolio responsibility are:

- Economics Committee – has broad-ranging policy scope and flexibility to capture all factors that enhance or impinge on farm businesses, including investment and taxation issues and drought policy.
- Trade Committee – is focused primarily on international trade and market access issues, World Trade Organization negotiations and free trade agreements.
- 2050 Committee – a new group dedicated to emerging issues and forward-looking opportunities for the farm sector, including long-term goals around,

for example, agricultural R&D, infrastructure and technological opportunities and challenges.
- Sustainability Committee – concerned principally with environmental considerations and maximizing the effective engagement of the farm sector with Environmental Stewardship, the Environmental Protection and Biodiversity Conservation Act and related issues.
- Water Committee – focused specifically on the ongoing importance of Australia's water reform agenda, its impact on farmers and regional communities and the need to ensure this most precious resource is managed sensibly.
- Workplace Relations Committee – covers industrial and workplace relations, education and training, skills development, human resource management and safety issues.
- Biosecurity and Animal Management Committee – oversight of quarantine and biosecurity issues to safeguard Australia from pests and diseases that could damage domestic farm production and Australia's international reputation as a source of clean, reliable and safe food and fibre, as well as animal care and husbandry practices.

As the need arises, the NFF also appoints taskforces to undertake investigation of policy options in relation to specific issues. The NFF currently has four active taskforces:

- Emissions Trading Taskforce – examines the application of greenhouse gas policies on Australian agriculture.
- Native Title Taskforce – ongoing monitoring and evaluation of native title and its implications for farm businesses.
- Drought Pilot Working Group – to engage with the federal government's proposed changes to drought policy, particularly the drought reform measures pilot.
- Mining and Coal Seam Gas Taskforce – to consider federal issues relating to the interface between agriculture, mining and coal seam gas, along with assisting member organizations to work collaboratively on state-based legislative matters.

In addition, the NFF also takes membership of various external committees. Through these external committees, the NFF provides input to the development, implementation and review of the policies and issues that guide and govern Australian farm practice. Currently, the NFF is represented on over 50 external committees.

A management committee oversees the operations of the NFF Secretariat, which is based in Canberra. The Secretariat comprises of one CEO, four general managers and three managers, plus several administrative supporting staff.

Since its creation in 1979, the NFF has acted as the single national voice for Australian farmers, leading a number of high-profile policy battles on issues such as workplace relations, tax, the environment and international trade reforms. Presented below are some areas where the NFF has exerted its influence in the policy process:

- **Free trade.** Achieving liberalization of the agricultural trading system has been one of the highest strategic priorities of the NFF since its inception, with

the federation leading the way as the first industry lobby in Australia to embrace and advance a free market philosophy. Continuing to lead on international trade reform, in 1998 the NFF initiated the Cairns Group Farm Leaders Forum – comprising farmer organizations from the 17 countries whose Trade Ministers formed the Cairns Group alliance in the mid-1980s – to press the cause of fostering world trade and bringing down trade barriers. Today, many Australian farmers accept that, in many cases, protectionist government policy would do them more harm than good and protectionist policy reduces the competitiveness of the farming sector.
- **Tax reform.** Another ongoing issue for the NFF is tax reform. From 'day one' the NFF has called on governments for broad-based tax reform and the removal of taxes on farmers' productive inputs. In the pursuit of this cause, the NFF has achieved several significant breakthroughs on behalf of farmers, which have bolstered the efficiency of the Australian economy.
- **Education and training.** Recognizing the importance of ongoing education and training in enhancing the competitiveness of Australian farming, the NFF has been instrumental in developing initiatives to deliver relevant, quality training and information to farmers.
- **The environment.** The NFF has led Australian farmers in being pivotal in the development of some of the most significant national environmental initiatives of the past 20 years. One of the most significant achievements in NRM came in 2003, when the NFF lobbied the Australian and state governments successfully for an Intergovernmental Agreement for a National Water Initiative, the aim being to provide environmental sustainability and long-term resource security for farmers, the environment and all Australians. Land and water resource security remains a strong focus of the NFF, with Australian farmers collectively investing billions of dollars of their own money, and time, each year sustainably managing environmental outcomes on behalf of future generations (see Section 9.2 for more details). In fact, Australian farmers plant over 20 million trees each year, solely for conservation purposes.

Today, the NFF is a key player in the major issues confronting modern Australia, including climate change and the development of an effective emissions trading scheme for Australia, water resource management, environmental stewardship and drought-proofing Australia – through appropriate drought management and preparedness.

The NFF has been regarded as a very successful farmer organization. The achievements of the NFF demonstrate that a united voice for Australian farmers is a powerful and extremely successful vehicle for agriculture to achieve substantial gains that benefit farmers – and Australia – economically, environmentally and socially.

4.2.3 Reasons for success

Australian farmer organizations have been very successful in providing support to individual farmers and protecting their interests. A number of reasons are responsible for such success:

1. **Willingness and dedication of farmer leaders.** Generally, those who assume a responsibility to run a farmer organization do their work on a voluntary basis. They receive little compensation, or compensation that is far from being comparable with the amount of time and energy they devote to their work (R. Sullivan, President of NTCA, NT, 4 December 2010; D. Gatenby, President of TFGA, TAS, 12 January 2011, personal communication). When asked why they are willing to dedicate their time, major reasons include: (i) passion for their industry; (ii) keen to make a contribution to their industry; and (iii) satisfaction from helping fellow farmers. Most of them serve their farmer organizations wholeheartedly without any hidden agenda for personal gains. None the less, some do move higher in the system (e.g. from local farmer organization to state or even national farmer bodies) to head for more important activities when they demonstrate their ability to lead, or some even move to state or federal parliaments. It is those unselfish and dedicated farmer leaders who drive the success of their organizations. This is the most important reason for the success of farmer organizations in Australia.
2. **Voluntary.** Whether or not to form a farmer organization is completely the choice of farmers. Whether or not a farmer decides to join a farmer organization is also at their own will. No one is forced to join a farmer organization.
3. **No government intervention.** There is no government intervention in farmer organizations in any way.
4. **Financial independence.** No financial support is provided by governments to any farmer organization and no such support would be accepted by any farmer organization anyway. Maintaining financial independence is important. Accepting government financial support invites government intervention and is the beginning of failure.
5. **Apolitical.** Being apolitical enables farmer organizations to be in a better position to negotiate with different political parties.
6. **Transparency of operations.** Farmer organizations are run with great transparency. They must maintain proper financial and membership records. Decisions that have a significant impact on the organization must be approved by the members. Executive members are required to declare whether they have any conflict of interest when important decisions are made. If a conflict of interest exists, then the member is excluded from the process of making the decision.

Entrepreneurial in the management of farmer organizations. Farmer organizations must be able to offer sound values to attract and retain members and associates. This is important for two reasons: to have a large support base and to have membership fee payments to sustain the organization. Some farm organizations are very innovative in attracting members by offering various membership categories and by providing more member services (e.g. the NSW Farmers' Association). Relying on membership is often not sufficient. Some organizations have been very innovative in generating extra funds to support their activities. One approach is to participate in bids for government-funded projects. Most state farmer organizations obtain such support for their activities; for example, the NTCA (L. Bowen, CEO of NTCA,

Darwin, 2 December 2010, personal communication) and the SAFF (C. Vincent, CEO of SAFF, Adelaide, 15 February 2011, personal communication). It is noted that when doing so, they are just like any other project fund applicant and there is no guarantee of their application's success. Such project funds are not government assistance for their farmer organization. Some have been business-like: they run commercial activities and earn some extra money to support their activities (e.g. the online shop of the NSW Farmers' Association). As noted earlier, the NSW Farmers' Association also invites some businesses to be associate members and achieves mutual benefits for such businesses and their members.

7. **Flexible.** Farmer organizations are formed for a purpose. The business environment, however, changes all the time. When major business environment changes occur, the earlier justification for a farmer organization may disappear. As such, the farmer organization has no more reason to exist and should disappear as well. The MMC Co-op (in Section 4.2.2) provides an example, showing that members are flexible in letting the MMC Co-op go when it is no longer justified to exist.
8. **Speak in one voice.** Following the creation of the NFF, farmers became more united. They have a strong voice in society. Their voice is generally heard – when there are important matters affecting the rural community, the NFF is generally consulted. Representing farmers of the country, the NFF has its inputs into government policy processes. It does not just wait passively to comment on or accept government policies, it also actively takes initiatives to lobby the government to change policies that have impacts on farmers.
9. **Self-assuredness.** Most Australian farmers have a relatively high education level and are very self-assured. They know the rights and obligations they have as citizens. They are well informed of market conditions and competition. Most of them accept that getting themselves organized is an effective way to defend their own interests. They do not wait for someone else in society to come to help them; instead, they help themselves.

4.3 Other Protective Measures

In addition to the efforts by farmers themselves, various other measures exist in Australia that are designed to ensure farmer interests are protected. The fundamental principle behind these measures is that everyone should be treated equally, no matter what one does and where one lives. Australians care about *fair go* and they defend this national spirit fiercely. Below are some demonstrations of how farmers are treated equally.

4.3.1 Healthcare

All farmers and their family members enjoy the same protection of Medicare (Australia's public health care programme) as any other citizens. For those rural residents who live a long distance from major centres and need to visit medical specialists, additional costs (costs that one would not have to incur if one lived in major urban centres, such as travelling and accommodation) are largely covered by Medicare. For those who live in very remote areas, the Royal Flying Doctor Services provide emergency treatment to patients, who are airlifted to a suitable hospital, and again additional costs are covered by Medicare.

4.3.2 Social security benefits

Anyone in rural areas is protected by the country's social security programme in the same way as their urban counterparts, such as unemployment payments, youth allowance, etc. This protection helps some farmers to overcome short-term, temporary or unforeseen financial hardship. But perhaps, more importantly, it greatly facilitates those who wish to exit farming: to have a less stressful and smoother transition when leaving the land.

4.3.3 Roads, electricity, telephone, postal and other communication services

Where practical, all such services are provided to rural residents. With the advent of the Internet and its wider adoption, Internet access is also made available to most of the farming community. Given the vast land mass of Australia, providing such services is obviously very costly. To some extent, cross-subsidization among users of such services takes place. For some rural families that live in very remote areas, household supplies and postal services may be provided by air delivery at a set frequency.

4.3.4 Equal voting power

In Australia, each of the rural voters has the same voting power as any urban voter. This may not be thought of as an issue; however, in some countries, farmers have less voting power compared with urban residents. For example, in China from 1953, soon after the People's Republic of China was founded, eight rural voters had the same voting power as one urban voter. In 1996, a change was made and four rural voters had voting power equal to one urban voter (Zhou *et al.*, 2010). A further change was made in 2010 in that rural and urban voters were to have equal voting power. The larger proportion of rural population should not have been a justification for rural voters to have less voting power. In India, the proportion of rural population was also very large at the time when the Republic of India was established in 1950. However, all rural adults have had the same voting power as their urban counterparts since 1950.

In Australia, not only do rural voters have equal voting power, but electorates were actually weighted in the past so as to benefit rural voters. The boundaries of electorates are drawn by independent commissions according to various guidelines. The most important guideline in the Australian history has been the balance between urban and rural electorates. An approach known as malapportionment, or unequal representation, was used to benefit rural voters. Malapportionment is the allocation of more electoral districts to one part of a country or state than its population would merit. This approach enabled electorates in rural areas to have fewer voters than urban electorates. Through this method, the National Party (previously the Country Party) received an advantage over their competitors.

The use of malapportionment was quite common in Australia in the past. There were several arguments defending the use of this approach: (i) country people had to contend with greater distances and hardships and thus deserved greater representation; (ii) country people (and specifically farmers) produced most of the nation's real wealth and thus deserved greater representation; and (iii) greater country representation was necessary to balance the radical tendencies of the urban population. In the later 20th century, these arguments were challenged and the use of malapportionment was reduced and finally abolished in all states. None the less, the bias in favour of rural voters by this approach played its historical role in fostering strong agricultural industries and rural development.

4.4 Concluding Comments

Australian farmers choose through elections who represents them politically in governments. Their representatives defend their interests at various levels of governments. If these representatives do not work to farmers' expectations, they will be voted out (where farmers are the largest voting block in an electorate). Australian farmers also form various organizations of their own to defend their interests. It is with strong political representation and their own organization that farmers in Australia have their opinions aired and their interests represented and protected in society. In addition, various other institutional arrangements and policies help ensure that farmers are treated equally and fairly. Farmers in Australia even received positive discrimination through the malapportionment approach in favour of rural voters.

Australian farmers have the ability to influence, both directly and indirectly, government policies that have an impact on their welfare. It is because of this ability that Australia's rural industries and various farmer issues are not neglected, in spite of the fact that today the relative importance of rural industries to GDP is small, being even less than 3%. The interests of the rural industries and the welfare of rural residents continue to be well looked after, to the extent that they accord with national welfare. This is a very important experience from which lessons can be learned by many developing countries where agriculture is dominant in the economy but poorly developed.

In many developing countries, the fact that farmers are still very poor is exactly because their interests and welfare are not represented fairly in society. Their concerns and voices are muted in society and cannot be channelled through to become inputs for policy formation. The literature shows again and again that farmers can be milked without immediate and significant consequences for government rulers who mistreat their farmers. This has been the case in many agrarian societies because there is a lack of political force that helps farmers to air their concerns and defend their interests and welfare.

In some countries, like the former USSR, agriculture has long been milked to generate funds to support urban development and industrialization. After the urban system was developed, rural industries were still left with limited opportunities to develop. Lack of farmer representation in society is the fundamental cause as to why many problems concerning agricultural development cannot be resolved.

In many developing countries, rural residents are treated unequally and unfairly. Many entitlements and social benefits available to urban dwellers are not available to rural people. Rural people often bear higher living and production costs due to lack of access to some essential services and infrastructures. In many such societies, agriculture as an industry, farming as a career and farmers as a group of very hard-working citizens are often looked down on by those who have forgotten that their ancestors came from the land. Forces representing farmers' interests are very scarce or weak.

Developing countries need to have forces that can represent farmers' interests and defend farmers' welfare. Such political forces should be allowed, encouraged and fostered. Without such representation, farmers cannot do well and rural areas cannot be developed in a way that is comparable with urban development. Agricultural development will be slow and will suffer, and ultimately the economy of the whole nation will suffer.

To learn from the Australian experience in protecting farmers' interests, several important aspects deserve attention:

1. Farmers should be treated as any other citizens, politically, economically and socially.
2. Access to services such as health care and social security, and to infrastructure for farmers, should be equitable.
3. Farmers establish and run their own organizations at their own choice but without government intervention.
4. Farmer organizations may choose to be apolitical.
5. Farmer organizations should refuse any direct financial support from any government departments to maintain their financial independence.
6. Last, and most importantly, those who decide to work for farmers' interest have to be prepared to provide voluntary but unpaid, or not fully paid, services to farmer organizations. They must not be selfish but willing to use their time and wisdom to help farmers. In some developing societies, it may be useful to foster a culture that encourages individuals to undertake voluntary work, generating benefits for others.

Notes

[1]The description of the origins of Australia's farmer organizations is based on Chislett (1967). There is a good amount of literature on the history and activities of farmer organizations in Australia. In all the three editions of the book *Agriculture in the Australian Economy* by Williams (1967, 1982, 1990), there is one chapter addressing farmer organization issues.

Connors (1996) provides useful discussions about farmer organizations in Australia, with a focus on how they became united to have one strong voice.
[2]Based on NFF (2010 and 2011) and RGA (2012).
[3]Based on NFF (2010 and 2011) and NSW Farmers' Association (2012).

References

Campbell, K.O. (1980) *Australian Agriculture: Reconciling Change and Tradition.* Longman Cheshire, Melbourne, Australia.
Chislett, G. D'A. (1967) Primary producer organisations. In: Williams, D.B. (ed.) *Agriculture in the Australian Economy.* Sydney University Press, Sydney, Australia, pp. 105–127.
Connors, T. (1996) *To Speak with One Voice: The Quest by Australian Farmers for Federal Unity.* National Farmers' Federation, Canberra.
NFF (2010) *Annual Review 2009–10.* National Farmers' Federation, Canberra.
NFF (2011) *Annual Review 2010–11.* National Farmers' Federation, Canberra.
NSW Farmers' Association (2012) Information about NSW Farmers' Association (http://www.nswfarmers.org.au, accessed 20 April 2012).
RGA [Ricegrowers' Association of Australia] (2012) About RGA (http://www.rga.org.au, accessed 15 April 2012).
Warhurst, J. (1990) The politics of rural Australia: rural industries and rural areas. In: Williams, D.B. (ed.) *Agriculture in the Australian Economy,* 3rd edn. Sydney University Press, Sydney, Australia, pp. 109–126.
Whiteley, W. and Zhou, Z.Y. (1998) Grass-root wheat marketing cooperatives in Australia: a case study. *Rural Economics Papers,* Ministry of Agriculture, China, No. 5, pp. 61–64.
Williams, D.B. (ed.) (1967) *Agriculture in the Australian Economy.* Sydney University Press, Sydney, Australia.
Williams, D.B. (ed.) (1982) *Agriculture in the Australian Economy,* 2nd edn. Sydney University Press, Sydney, Australia.
Williams, D.B. (ed.) (1990) *Agriculture in the Australian Economy,* 3rd edn. Sydney University Press, Sydney, Australia.
Zhou, T.Y., Cui, Q.X. and Jia, N. (2010) Points to pay attention in the new election act (http://news.xinhuanet.com/politics/2010 03/11/content_13169111.htm, accessed 15 November 2010).
Zhou, Z.Y. (2004) China's experience with agricultural cooperatives in the era of economic reforms. *China Agricultural Economic Review* 2(2), 238–257.

5 Deregulating Agriculture

The substantial deregulation of agriculture by the Australian Government has to be one of the most celebrated achievements in Australian agricultural development. Australia's agriculture used to be heavily regulated. Agricultural economists in Australia had long argued for deregulation, but the pace of change quickened only after the release of the Green Paper into agriculture, commissioned by the Whitlam Labor government in 1974.[1] Since then, support to rural industries has been gradually reduced or withdrawn. Deregulation required farmers to make their own production and marketing decisions, and fostered their capability to handle the changing business environment. It made Australia's agricultural sector internationally competitive. This chapter shows the experience of agricultural deregulation in Australia.

In the next section, a brief history about the extent of agricultural regulation prior to the 1980s is presented. This helps the reader to appreciate how daunting the task was. Much of this section is based on Mauldon (1990), with reference to Lewis (1967) and Lloyd (1982). Section 5.2 demonstrates how the deregulation of agriculture was carried out, with examples from several major industries. Section 5.3 explains the approaches used by the Australian governments to win over the cooperation and support of industries for the deregulation and to provide due assistance to help the farming community to adjust to changes caused by the removal of regulations. Section 5.4 highlights how the role of government in agriculture has evolved following deregulation. Section 5.5 summarizes Australia's experiences in agricultural deregulation and discusses how such experiences might be valuable to other developed and developing countries.

5.1 Heavily Regulated Agriculture Prior to the 1980s

The start of agricultural regulation can be traced back to the beginning of federation or, if the truth be known, back to early settlement days. Since then, a variety of price policy instruments or schemes had been used to regulate agriculture. These instruments or schemes generally were used for multiple purposes, with varying emphases placed on regulating marketing, providing assistance to producers and reducing price fluctuations and uncertainties. From the mid-1950s to the early 1970s, considerable income was transferred to farmers from domestic consumers through home consumption pricing. In the early years of this period, these transfers were seen as part of a national strategy for encouraging exports in order to finance industrialization and population growth, with full employment through economic diversification into manufacturing (Mauldon, 1990).

5.1.1 The start of Commonwealth intervention

Commonwealth intervention in pricing commenced at the time of federation. This intervention was introduced to guarantee QLD a profitable market for its sugar in order to lure QLD to join the Commonwealth. The QLD sugar industry was founded in the 1860s. During the 1890s, there was a move to replace previously large estates with small farms based on European labour. This required protection of the industry against imports. In 1902, the new Commonwealth imposed an excise on manufactured sugar and paid most of the collected money as a bounty on cane grown and harvested by European labour. This supported a domestic price at a level that was comparable with the landed, duty-paid price of imported sugar.

Prior to 1923, Australia was an importer of sugar. In 1923, Australia became an exporter. The Commonwealth and QLD governments entered into an agreement under which the Commonwealth prohibited the importation of sugar and QLD controlled prices on the Australian market, at levels agreed by both the Commonwealth and QLD governments, by compulsorily acquiring all sugar grown in QLD (and purchasing the small amount grown in NSW). This agreement was renegotiated from time to time, with these basic features remaining until the agreement finally lapsed in 1989. This home consumption price originated as compensation for a labour-cost disability imposed by federation, but it was impossible to quarantine the principle from spreading to other industries, though for different reasons.

5.1.2 Marketing crises of the 1920s

Towards the end of the First World War, world prices for rural products boomed. This led to rapid expansion of production in Australia, with exportable surpluses. However, in the early 1920s, a number of marketing crises caused the Commonwealth government to intervene in order to prevent ex-soldier settlers from leaving their recently established farms, and to defend other public investments, particularly irrigation. Initially, bounty payments were made to offset what was thought to be temporary losses on some fruit marketing pools. However, the government later sought solutions that became more permanent and that placed fewer onuses on government expenditure.

These solutions were designed to reduce the variability in farm prices by controlling the domestic price (often based on some notion of cost of production) and diverting surpluses to lower-priced export markets. For this purpose, the Commonwealth government placed export sales of several commodities under the control of export control boards. These were established for dried fruits and dairy products in 1924, canned fruits in 1926 and wine in 1929. However, processors would only divert supplies to export markets if returns from them were profitable. The government sought to finance incentives to export by maintaining domestic prices at higher levels (which required stocks to be exported) than export prices and remitting a uniform price to growers from all sales often based on a notion of the cost of production. These efforts were successful initially for dried vine fruits, butter and canned fruits, in addition to sugar. Some of these initiatives complemented state orderly marketing arrangements, which had been introduced a few years earlier to control interstate sales.

5.1.3 Discriminatory pricing within domestic markets

Price discrimination policies for different domestic uses of an agricultural product were first introduced for the dairy industry. Milk produced for liquid consumption had been controlled by the states for health reasons since the early years of federation. These controls raised the production costs of milk for this use; however, its price tended to be determined by the value of milk in other uses. Consequently, supplies and prices for the liquid milk market were erratic. Since milk for liquid consumption had staple food status, the states established milk boards in the 1930s and 1940s to control the production, distribution and pricing of liquid milk sold in metropolitan areas in order to stabilize supplies and consumption. Higher prices were offered for milk for fresh liquid consumption to offset the costs imposed by health standards and requirements to supply continuously throughout the year. Freight costs provided sufficient natural protection to prevent price erosion from interstate trade. The implementation of such discriminatory policies also required quotas to be set that would entitle farmers to sell a certain amount on the high-priced liquid milk market.

Discriminatory pricing within domestic markets was also practised for sugar, with lower prices for use in brewing, for which sugar faced a more elastic demand as a fermentable than as a sweetener. It was also practised for wheat, with lower prices for industrial use and as stockfeed, for which wheat faced more elastic demand than for use in human consumption.

5.1.4 Wheat policy in the 1930s

A 'grow more wheat' campaign was introduced in 1930. The world wheat price was sluggish in the early 1930s, from which Australian wheat producers suffered.

The wheat industry subsequently became the subject of a Royal Commission from 1934 to 1936, which for the first time subjected home consumption pricing to independent public scrutiny. The Royal Commission recommended that a home consumption price should operate as a relief measure for the limited periods of low world prices, enabling wheat growers to share the benefits accorded other industries by the protection policy of the nation.

However, when legislation was introduced in 1938 for a home consumption price for wheat used for human consumption, there was no suggestion of it being a short-term relief measure. Thus, by the end of the 1930s, home consumption prices were being justified as part of a philosophy of 'protection all round'. The rationale was that, since labour was protected by arbitration, manufacturers needed to be protected by tariffs, and to enable farmers to share the benefits of an industrially diversified economy, they should be able to charge Australian consumers higher-than-export or import-equivalent prices. This doctrine was embraced by the wartime Rural Reconstruction Commission.

5.1.5 Continuing bounty payments

Before the Second World War, prices of some farm products had been supported from time to time by bounty payments. Some of these bounties were financed from excises imposed on processors, which raised domestic prices. Others were financed directly from general revenue, without imposing direct costs on domestic consumers. The earliest directly funded bounties were introduced in 1907 for a number of minor import-substituting commodities (e.g. vegetable fibres, oilseeds, rice, coffee, tobacco and some dried fruits), as part of the 'new protection' policy of that year. These bounties operated for 5–15 years, but little production was induced by them and little was paid out.

Several direct bounty payments were also made to fruit marketing pools in the 1920s and for wheat in the early 1930s. In contrast to the earlier bounties, these later bounties supported export-oriented activities and were enacted in an ad hoc manner from year to year. During the Second World War, bounty arrangements emerged for manufactured dairy products. Under special wartime powers, the Commonwealth government set home consumption prices for butter and cheese as part of general domestic price controls. In the early war years, production costs increased rapidly, causing a sharp reduction in dairy farmers' incomes. In order to maintain exports of dairy products, without jeopardizing price and wage stability, the government provided bounties to manufacturers of butter and cheese in lieu of a domestic price rise. These were paid to manufacturers on condition that they participated in equalization arrangements designed to pass the bounties back to farmers. The bounty payments continued on an annual or biennial basis until 1947, when they were extended for 5 years (dependent on price outcomes) as a means of guaranteeing dairy farmers a return that would cover the assessed average costs of production.

By the 1950s, the bounties had become a pivotal element underpinning the price equalization arrangements. They were continued in the next 5-year plan, but the basis of payment was determined annually by the Commonwealth. By the mid-1950s, they were set as a fixed sum, independent of price outcomes. In 1973, bounty payments were phased out over 2 years.

5.1.6 Price stabilization schemes

Marketing arrangements for manufactured dairy products during the 1940s also encompassed other features designed to raise and stabilize farmers' prices and equalize returns. Some of these had operated since the early 1930s. Statutory boards in each state set quotas for the amount of the state's production which could be sold on the higher-priced domestic market within the state, thereby diverting surplus production to exports or interstate sales. However, a condition for receiving the bounty payments was participation in the equalization arrangements. Alternative sources of satisfying domestic demand were also controlled, through state quotas on the production of table margarine and a virtual embargo on imports through tariffs and import licensing.

Some other features of wartime controls also gave impetus to more formal post-war pricing and marketing arrangements. At the beginning of the war, the Commonwealth government negotiated a number of bilateral agreements under which the UK purchased Australia's exportable surpluses of bulk agricultural commodities. These agreements were continued until the early 1950s. The wartime Rural Reconstruction Commission recommended a formal framework for price stabilization incorporating the sorts of features which were operating for dairying. Price stabilization subsequently was endorsed as a component of the Commonwealth government policy for agriculture.

The Commonwealth government hoped that the wheat and dairy industries would set up funds to transfer returns from years when export prices were high to years when they were low. This became a feature of minimum price guarantee arrangements for butter and cheese from 1947 to 1952. It also became part of a statutory price stabilization scheme for wheat in 1948. Other statutory schemes embodying discriminatory pricing, equalization and stabilization funds were established for dried vine fruit in 1964 and for apples and pears in 1971. These schemes were referred to as buffer fund schemes. A statutory scheme that did not involve a stabilization fund was established for eggs in 1965. The price stabilization schemes for wheat and eggs are worth particular mention, as follows.

The stabilization scheme for wheat commenced in 1948 and, with modifications, operated until 1989. Wheat stabilization plans were renegotiated about every 5 years. Under each plan, the AWB was given control of wheat sales on both the domestic and export markets. This was achieved through the board's compulsory acquisition powers, which were underpinned by complementary state and Commonwealth legislation. Domestic sales were at administered prices, with the exception of some stockfeed wheat from 1984 to 1989. A pool was established for each season's wheat, and growers received an equalized price from domestic and export sales.

It was arguable whether the domestic pricing mechanism in the wheat stabilization scheme was intended unequivocally to assist wheat growers, since the assistance actually received during each of the plans could differ from the assistance thought likely at the time the plans were negotiated. In periods of unanticipated high world prices, such as occurred from 1973 to 1975, the home consumption price severely taxed wheat growers. In the plans that operated until 1984–1985, a new domestic price was negotiated at the beginning of each plan. This price was adjusted annually in relation to movements in the assessed costs of production. This had the effect of protecting wheat growers to some extent against cost increases that were beyond their control.

The wheat plan was not so much about assistance, but about tempering price variability and gaining some monopoly benefits from a single desk seller. However, the guaranteed price was difficult to adjust downwards, partly because it was related to the cost of production and partly for 'political reasons'. Farmers were basing their production decisions on the equalized price rather than on the often lower world price, hence building in incentive for excess production.

A national egg equalization scheme is another example. It existed from 1965 to 1987. Prior to 1965, there was no single national price for eggs in Australia. Each state had an egg marketing board whose policy was to ensure a sufficient supply of eggs for its domestic market at prices which the board determined. At these prices, more eggs were produced than could be sold profitably. Prior to the national egg equalization scheme, some producers had incentives to sell eggs interstate outside of the marketing boards and thus avoid having their returns reduced from export sales.

In 1965, the Commonwealth government imposed a tax on laying hens to be paid through the egg marketing board in the state where the eggs were produced. The money collected was pooled nationally and an equalization payment made to each of the boards in relation to its exports. This supported domestic prices above export returns by an amount equal to the equalization payment. Producers received an equalized price for all eggs sold through the board. The size of the tax on laying hens and the size of its recoupment from the board were sufficient to induce compliance. However, at the domestic prices chosen, there were incentives to produce considerably more eggs than were required for the domestic market. Thus, the boards imposed state hen quotas to ensure that surpluses were small.

5.1.7 A local content scheme for tobacco

Price discrimination was not a feature of tobacco, since there was no surplus product to export. Tobacco has been grown in Australia since the 1860s, but domestic supplies have never been adequate to meet the range of domestic needs. In 1936, the Commonwealth government sought to encourage greater use of Australian leaf by introducing a local content scheme under which tariff concessions were provided to local manufacturers who used no less than a minimum specified proportion of Australian leaf in their tobacco products.

The initial percentage requirements were modest, but they were increased steadily from the mid-1950s and had risen to 57% by 1977. In response, tobacco growing increased rapidly from the mid-1950s, resulting in difficulties in disposing of the crop during the early 1960s, even though large amounts of tobacco were still being imported. In 1965, the Commonwealth government introduced a stabilization plan that retained the tariff concessions for manufacturers who used specified minimum proportions of Australian leaf. Supplies were managed by allocating marketing quotas to growers through tobacco leaf marketing boards in each of the growing states. The Australian Tobacco Board constructed minimum price schedules and allocated aggregate state marketing quotas.

5.1.8 A buffer stock scheme for wool

Despite the growth of price stabilization arrangements during the 1950s and 1960s, not all (or even most) agricultural industries sought protection through price intervention. Wool, meats, feedgrains, most vegetables, many fruits and products of small industries such as honey received little price support based on statutory measures. Wool was sold in a decentralized manner and very little was processed within Australia. This precluded home consumption pricing as a means of price stabilization or support. Also, the wool industry was relatively large and politically powerful and there was a determination to remain independent of government intervention.

However, Australia produced a large share of the world's fine wool and it was perceived to be advantageous for Australia to limit the amount of wool produced or to control the flow of supplies into the market. These issues were debated during the 1950s and 1960s. Following a recession in wool prices during the late 1960s, the Commonwealth government intervened. In 1970, it empowered the Australian Wool Commission to seek to reduce short-term fluctuations in auction prices by buying wool which failed to meet a 'flexible reserve price' and subsequently selling it when prices improved. Stocks of wool initially were financed by the government and from trading bank loans. In 1974, a 'minimum reserve price' was introduced to provide growers with a guaranteed minimum price for their wool. The scheme was funded by a proportion of the tax paid by growers on the value of shorn wool, and was administered by the Australian Wool Corporation (AWC), which purchased all wool not meeting the minimum reserve price at auction. This wool would then be sold later during periods of higher prices. Thus, a buffer stock was built up from which wool could be sold when prices were satisfactory.

5.1.9 Price discrimination between export markets

Although Australia had limited global power for rural products other than wool, it was still possible to extract price premiums for some products in particular markets because of their proximity or because of goodwill to Australian suppliers. Several statutory marketing authorities with export control powers had sought to prevent the erosion of potential price premiums that could result from competition among Australian exporters. This was straightforward where a commodity, such as wheat, was acquired compulsorily and exports could be directed to specific markets. Where this was not the case, the authorities often limited the number of export licences and restricted the activities of exporters by setting minimum export prices or directing sales to specified agents.

5.1.10 Subsidies on agricultural inputs

Diverse agricultural products are produced in rural Australia. It would be very difficult to assist all of them through output prices. Farmers and governments were attracted to interventions in a small number of significant inputs used by many farmers in a wide range of industries. The two most generally used input subsidies were for finance and fertilizers. Irrigation water was also provided by state governments at less than cost. Other forms of input subsidies included tax concessions on land clearing, fencing and pasture improvement, and losses on state rail services.

Subsidies on the use of fertilizers are an example of the persistence of some pricing interventions. What was introduced as a relief measure in 1932 (except for wheat growers, for whom other measures applied) was transformed into a measure to help control consumer prices during the Second World War and lower farmers' costs in the early post-war years. Subsidies were retained for nitrogenous fertilizers to the mid-1950s to compensate fertilizer-using industries for not being covered by price stabilization arrangements. The subsidies lapsed in the latter half of the 1950s but were reinstated in the mid-1960s to improve agricultural productivity and growth. Fertilizer subsidies were removed in 1988 as part of wide-ranging reductions in assistance to industries.

5.1.11 Rural reconstruction and adjustment in the early 1970s

While the agricultural price policies in the 1950s and 1960s were successful in expanding Australia's agricultural exports and earning foreign exchange to finance Australia's rapid industrialization and population growth, not all farmers managed to increase their income. For example, in the case of the dairy industry, a large number of dairy farmers were earning unacceptably low incomes in the late 1950s, in spite of rising levels of production, increasing productivity, domestic price support and production bounty payments and protection against imports. By the mid-1960s, a large number of low-income farmers emerged in most rural industries and in most regions.

Agricultural pricing and marketing policies were partly responsible for the development of these problems. They discouraged necessary rural adjustments to deteriorating terms of trade. Very little of the assistance that was being provided through pricing policies went to support low-income farmers, or to finance their adjustments in order to survive or exit from farming. In recognition of this, the Commonwealth government introduced a number of reconstruction schemes in the early 1970s. These schemes departed from the ways in which assistance to agriculture had been delivered in the 1950s and 1960s. Instead of focusing on prices, they focused on flows of funds, on the structure of assets and liabilities and on impediments to the mobility of people and capital. Instead of being product based, they were farm business based. Their objective was to assist small and inefficient farmers to leave agriculture.

5.1.12 Public scrutiny of agricultural policies

Unfortunately, the reconstruction schemes of the early 1970s augmented rather than replaced the policies of the previous two decades. A more thorough redirection of policy seemed to be impossible under the Liberal–Country Party coalition government, which had been in power, unbroken, for almost a quarter of a century. In late 1972, Labor came to power. Following a review commissioned by the new Labor government of the agricultural assistance policies of the previous government, the dairy product bounties were phased out. In 1974, the Labor government abolished the Tariff Board, which had existed since 1921, and established the IAC to extend the system of public scrutiny to all industries, including agriculture. It also commissioned a working group to prepare a discussion paper on principles for rural policy in Australia. Its report (the Green Paper) was the first public review of overall rural policy sponsored by the Commonwealth government in nearly 30 years.[2]

Soon after the publication of the Green Paper, the IAC sought and received a reference on the scope for reducing fluctuations in the incomes of rural producers. In 1977, the Income Equalization Deposits Scheme (operating through the taxation system) and carry-on loans and household support as part of the new RAS were introduced. By then, the stage was set to remove regulations and controls over agriculture.

The above discussion highlighted that, prior to the 1980s, Australian agriculture was heavily regulated, chiefly through price policies. The instruments and schemes used were very diverse; some major ones are listed below:[3]

- price discrimination;
- import protection;
- export subsidies;
- output subsidies;
- input subsidies;
- market-share quotas;
- export controls;
- local content;
- pooling;
- price and income stabilization;
- buffer stocks; and
- buffer funds.

The diversity of Australia's regulation measures compared with other countries is very striking. According to Lloyd (1982), Australia used most of the many cards in the policy pack. The diversity was due to differing circumstances and needs, over time and between industries. Also contributing to the diversity was the nature of Australian federalism, under which the states had constitutional responsibility for agricultural production and for the fixing of prices for all products, including rural products (see Section 3.1). To impose a comprehensive and uniform national price policy for farm products was very difficult because of the virtual veto power of state governments, whose interests would frequently conflict because of historically and geographically based differences.

Mauldon (1990) noted that while the diverse policy instruments and schemes used were successful in meeting many of their goals, they also resulted in many problems including, for example, over or under production, delayed adjustments, welfare transfers from consumers to producers and inefficient resource use. In hindsight, policies promoting agricultural development contributed to the growing impact of agriculture on the environment. More rigorous agricultural deregulation was thus called for.

5.2 Completely Deregulating Agriculture (since the 1980s)

Deregulation is the removal or simplification of government rules and regulations that constrain the operation of market forces. Since the early 1980s, Australian governments at both the federal and state levels have been proactive in deregulating many industries, including agriculture. Economists in Australia played an important role in addressing the economic problems caused by unnecessary intervention and in advocating the benefits of deregulation. Over the past three decades, deregulation has been carried out in almost all industries, making Australia one of the least protected economies in the world.

Australia's programme of economic deregulation gathered momentum from 1983. Controls on the domestic financial system were relaxed and competition was fostered; foreign exchange controls were abolished; international trade barriers were reduced and import quotas abolished; the growth of the shares of government revenue and expenditure in GDP was halted and even reversed; competition was introduced in telecommunications and transport; restrictive labour practices were reduced; and government business enterprises were corporatized or privatized.

In relation to agriculture, substantial efforts were made to deregulate this industry. In the meantime, agriculture benefited from the reforms and deregulations in the broader economy. As a result of 30 years' efforts, today, nearly all regulations concerning agricultural pricing and marketing have been removed for all rural industries, making Australian agriculture one of the least protected.

In the rest of this section, the processes of deregulation in several major agricultural industries in Australia are highlighted. These include: wheat, dairy, eggs, wool and sugar.

5.2.1 Wheat

The first major deregulation took place in the domestic marketing of wheat. In the 1970s and 1980s, domestic prices were related increasingly to export prices. For the wheat marketing plans that commenced in 1968 and 1974, the initial home consumption prices were negotiated according to the outlook for world wheat prices at those times rather than in terms of production costs. However, the initial domestic prices continued to be adjusted annually in terms of movements in assessed costs of selected inputs. In the plan beginning in 1968, the AWB was empowered to sell wheat on the domestic market for industrial and stockfeed uses at lower prices than for less price-sensitive human consumption use. In the plan beginning in 1974, there was no differentiation according to use.

For the plan that commenced in 1979, domestic prices again differed according to use. Domestic prices of wheat for industrial and stockfeed uses were set at the board's discretion, but generally in terms of export returns. The domestic price of wheat for human consumption was adjusted not only for cost changes but also for changes in export prices, being maintained about 20% above the export price. In 1984, the domestic stockfeed market for wheat was deregulated. In 1989, the domestic wheat market was fully deregulated. This reduces the burden of the government to contribute to the buffer fund. As noted earlier, prices were regularly set above the world price. In the meantime, growers were attempting not only to stabilize prices but also to stabilize them upwards. The cost to the government was increasing. Eventually, the buffer fund scheme was no longer sustainable.

Wheat exports were still under the control of the AWB. This single desk export arrangement was kept with a view that it would help a coordinated marketing effort in the world market to enable growers to achieve better export prices – it was about retaining market power against the large world grain trading companies and single desk buyers in some importing countries. Most growers, especially smaller ones, supported this arrangement. Some growers, especially those larger ones in WA, argued that the industry would be better off operating in competitive markets because there was little evidence that the AWB was extracting price premiums. They preferred that they could make their own arrangements to export their wheat. Whether or not the single desk arrangement should be kept had been subject to debate since 1989.

The AWB ceased operation in 1999 as a government-controlled statutory authority. It became AWB Limited, a grower-owned and -controlled corporation. At this time, all government financial assistance, such as the underwriting of borrowing, ceased. Through a subsidiary company, AWB (International) Limited (AWBI), AWB Limited continued to be the sole exporter of bulk wheat from Australia via the 'single desk' system. The export of wheat in containers and bags was allowed, but subject to the control of the Wheat Export Authority, the government regulator. By July 2007, container and bag exports were completely deregulated (WEA, 2011).

In 2008, the Australian Government removed AWBI's monopoly on bulk wheat exports. The single desk wheat export arrangement was finally abolished. However, the bulk export of wheat is still subject to some control from the government at the time of writing this chapter. That is, those who wish to export wheat in bulk have to get a permit from Wheat Exports Australia, which was established in 2008. In 2010, the PC report (PC, 2010) recommended the abolition of all the remaining control measures, except that the port access test was to continue until 2014. The Commonwealth government agreed in principle with the PC's recommendations, but indicated that it would delay abolishing the Wheat Export Accreditation Scheme and the Wheat Export Charge until 30 September 2012. Wheat Exports Australia would cease to operate by 31 December 2012 (WEA, 2011). From 1 January 2013, Australia's wheat markets will be fully deregulated.

5.2.2 Dairy industry

As shown earlier, the dairy industry in Australia had a long history of price support. For many years it gained assistance from support policies that allowed farmers with quotas to earn higher returns on sales in the domestic market. Two stages of reforms were used in deregulating the dairy industry, each with different approaches. The first stage was a lengthy, phased reduction in assistance for the export sector, which commenced in July 1986. The second stage involved the overnight elimination of all price support mechanisms in July 2000 – the elimination of quotas and the removal of restrictions on trade between states. The decision was announced 9 months before it was implemented. A detailed case study of dairy industry deregulation is given by Harris (2005a). Harris (2005b) also provides a shorter version of the reform process of the dairy industry, together with the reform processes of three other industries, where different adjustment approaches have been used.

5.2.3 Eggs

The egg industry was subject to a high degree of intervention through both state and federal marketing boards. Both production control and price discrimination were used. Supply control was enforced through quotas on laying hens. Prices were determined according to end use (fresh consumption versus manufacturing) and destination (domestic versus export). The control over egg production and marketing caused numerous conflicts between egg producers and state marketing boards. According to Piggott (1990), 'The New South Wales Egg Corporation and its predecessor often found themselves in a litigious state because of rebel egg producers selling their eggs outside the Corporation which, like most other State marketing boards and corporations, had vesting powers. In 1989, the NSW government abandoned the statutory marketing of eggs. Egg producers received up to AUS$16/bird for their loss of the quota asset. Following the NSW deregulation, other states started to deregulate their egg industry. SA was the second state to deregulate its egg industry in 1992. Unlike their NSW counterparts, who received a AUS$61m assistance package following their 1989 deregulation, SA producers did not receive anything. The last state to deregulate its egg industry was WA, where the egg industry was deregulated in July 2005.

5.2.4 Wool

For about 20 years, the wool reserve price scheme worked well. By the early 1990s, a massive wool stockpile had accumulated. Two major factors contributed to the huge stockpile. One was a sharp fall in demand for wool and the other was the high reserve price, which was set during a period of high demand in the late 1980s and was not adjusted downwards in line with the market. The high stockpile led to the scheme being suspended in February 1991, when the size of the AWC stockpile reached 4.7 million bales. The federal government, with the agreement of the industry, decided that the scheme could no longer be maintained. Australian wool producers went through a very painful adjustment process.

Initially, the Australian Wool Realization Commission was responsible for the disposal of the wool stockpile. In December 1993, the disposal of the stockpile became the responsibility of Wool International (WI), a statutory corporation of the Commonwealth government. WI was required to sell the stockpile in accordance with a statutorily imposed disposal schedule, the last bale of stockpile wool to be disposed of by 31 December 2000. By 30 June 1998, under the management of WI, the stockpile had been reduced to 1.2 million bales. By October 1998, equity in the wool stockpile had reached a level significantly higher than the wool debt and, therefore, ongoing government involvement in stockpile management was no longer justified.

On 15 October 1998, the Commonwealth government announced a freeze on sales of wool from the stockpile, and an intention to privatize WI by 1 July 1999. On this date, WI became WoolStock Australia Limited, a public company limited by shares allocated to previous holders of units of equity in WI. WoolStock Australia took over the assets and liabilities from WI and was fully accountable to its shareholders for the efficient management and sale of the stockpile. The principal activities of WoolStock were selling the stockpile and making distributions to unit/shareholders. By 9 August 2001, the stockpile was sold completely (ABS, 2002, p. 478).

Following deregulation in the early 2000s, the Australian wool industry continued to suffer from falling demand for natural fibres and a decrease in wool prices worldwide. Producers have adjusted to the changing market conditions by reducing their wool sheep numbers further and shifting to producing more meat sheep, taking advantage of the rising price of lamb in recent years. Some of them have moved out of wool production to produce other products or opted to have more mixed enterprises with the ability to produce more wool should profitability justify this in the future.

5.2.5 Sugar

By 1983, the sugar industry was still the most regulated rural industry in Australia, cementing in place many political interests and outmoded practices (Mauldon, 2009). In that year, the IAC was directed to review assistance to, and regulation of, the sugar industry. The IAC in its 1983 inquiry recommended changes that would expose the industry to greater forces of market competition. Little immediate change to the industry's regulatory framework emerged. But later in the decade, following further evidence of regulatory costs from the then Bureau of Agricultural Economics, some of the worst aspects of land area controls were relaxed and the level of price support under the Commonwealth/QLD sugar agreements was reduced. The embargo on sugar imports was replaced with a specific-rate tariff, which introduced a degree of import competition into domestic pricing.

However, most of the interventions supporting growers and millers depended on QLD legislation, and the structure of these interventions remained largely intact. Moreover, although assistance to most rural industries had declined by the beginning of the 1990s, it remained high for sugar. In view of this, in 1991 the

Commonwealth government again referred the regulatory and assistance framework of the sugar industry to the IC (the IAC became the IC in 1990).

The IC found that the major factors amenable to government actions which were impeding the development and efficiency of the Australian sugar industry stemmed from the regulatory controls applying to the production and marketing of raw sugar. It recommended the immediate cessation of compulsory acquisition of all raw sugar by the Queensland Sugar Corporation (QSC) (with some transitional arrangements to cover the QSC's forward contracts) and, after an initial expansion of assignment areas, the phased abolition of all constraints on land used for cane production. It also proposed the abolition of constraints on mills to which cane could be delivered, and that growers be able to negotiate contracts with millers on cane delivery arrangements and on the distribution of revenue from raw sugar sales. The IC also proposed that tariff protection applying to sugar imports, which impacted on about 20% of industry sales, be phased out and that transitional payments be made by the Commonwealth government to growers and millers as compensation during that time frame.

These proposals, however, were largely put off by both the Commonwealth and QLD governments (Mauldon, 2009). Sugar interests, particularly those in traditional growing areas, were powerful in QLD state politics. Several of the Commonwealth government's parliamentary members also narrowly held their seats in traditional sugar growing regions. This meant that immediate political considerations inevitably took precedence over longer-term efficiency and social adjustment issues in decisions made by the relevant governments. Thus, despite the IC inquiry, the status quo in the sugar industry remained. The Commonwealth government did inject AUS$40m into infrastructure works for the sugar industry. This, however, was additional assistance rather than, as proposed by the IC, compensation for removing tariff protection.

After the IC's 1991–1992 sugar industry inquiry, many changes occurred in the regulatory environment in Australia. As time passed, it became increasingly apparent to both the Commonwealth and QLD governments that a fundamental restructuring of the sugar industry's institutional arrangements was required. Both the Commonwealth and QLD governments provided significant packages of financial assistance to help restructure the industry.

In 2005, the QLD government abolished the cane area entitlement system and gave growers greater choice in how to bargain with millers. The industry later agreed to replace compulsory acquisition marketing with voluntary arrangements. In 2006, the QSC became a contractually based marketing company competing with other traders. Thus, 14 years later, the changes recommended by the IC in its 1992 report finally came to reality. The sugar industry had been the first to gain protection but was one of the last to have its protection removed.

5.3 Providing Adjustment Assistance

Deregulation in Australia did not occur without resistance. In the case of wheat market deregulation, in 1989 when domestic marketing restrictions were removed and in 2008 when AWB export single desk was abolished, not all farmers were happy and receptive to the changes. While marketing tools like forward contracting and futures markets were available, the highly uncertain climate (and hence production) made their use by many farmers difficult.

The deregulation of the sugar industry represents another typical example showing how industry resists change. Mauldon (2009) provides a detailed record on the process of the sugar industry deregulation from the early 1980s. For about 30 years after the then IAC made the recommendation to deregulate, the sugar industry strongly resisted any of the changes proposed by every inquiry. Considerable losses were imposed on the industry, and the nation, from opportunities missed along the way.

Given that rural industry adjustment is a painful process, farmers were generally provided with assistance to help them to adjust. Sometimes, the provision of such assistance was in the form of compensation for reduced asset values (where quotas were capitalized in land or livestock values, for example), in order to gain cooperation from the industry. Other types of transitional assistance, including training in risk management tools and concepts, were also used by governments, depending on the particular industry's circumstances. In this section, assistance policies adopted by the Australian governments over the past three decades are highlighted.

The Australian Government has used a variety of rural adjustment-type schemes since the introduction of a debt reconstruction scheme in 1935. Similar schemes continued until the late 1960s. In the 1970s, the Rural Reconstruction Scheme (RRS) changed focus in response to the structural adjustment pressures facing the farming sector and, in 1977, it was replaced by the RAS. The RAS continued until 1997, with modifications in 1985, 1988 and 1992. In 1997, the RAS scheme was replaced by the *Agriculture – Advancing Australia* (AAA) rural policy package. In 2008, the Labor government released a new package, *Australia's Farming Future.*

5.3.1 Rural Reconstruction Scheme (1971–1976)

The RRS came to effect in 1971. It allowed for debt reconstruction and farm build-up. It also contained a programme for 'rehabilitation' grants, which was a direct inducement for farmers to leave agricultural industries. More than 21,000 applications for assistance in the various programme categories were received from 1971 to 1976, but there were only 300 applicants for the rehabilitation grants, of which 197 were approved. The maximum of AUS$2700 seemed too little incentive to leave the land compared with the rewards for the successful applicants for debt reconstruction, who received an average of AUS$37,000, while those eligible for farm build-up received an average of AUS$46,000. Alternatively, perhaps the scheme provided an opportunity to apply for 'soft' finance and farmers responded accordingly. The RRS did little to reduce the number of farmers directly (Cockfield and Botterill, 2006).

5.3.2 Rural Adjustment Scheme (1977–1997)

The RAS replaced the RRS in 1977. All major strands of the RRS were incorporated into the RAS. The provision of 'soft' finance remained, through the programmes for debt reconstruction, farm build-up and farm development, but the criterion of 'long-term viability' applied. Carry-on loans were available for specific areas/industries, but only on a short-term basis. So, the restrictions on finance increased slightly, as did the incentives for industry exit. Rehabilitation loans (convertible to grants) were set at a maximum of AUS$5000 and a new welfare initiative, Household Support, was set at a maximum of AUS$3000, amended to AUS$5000 in 1979 (Cockfield and Botterill, 2006). The emphasis in this scheme was on saving the viable farmers and helping non-viable farmers to leave the industry.

The RAS went through three amendments. Following a review of the RAS, some amendments were made to the scheme in 1985. The carry-on programme was dropped, on the grounds that this was a straight 'welfare' programme that could be handled within the economy-wide welfare system. Rehabilitation assistance was increased to AUS$8000. Further, although the federal government was still to provide concessional loans via the states, the emphasis shifted to subsidizing the interest on commercial loans. The goals of the 1985 scheme had shifted subtly from maintaining viable farmers to a means by which the government augmented the capital market in encouraging the process of capital formation and adjustment in the rural sector (Cockfield and Botterill, 2006).

The subsequent review of the RAS in 1988 suggested that the RAS in 1985 allowed unviable farmers to delay off-farm adjustment decisions too long, neglected other aspects of adjustment, such as training, and spread the money too broadly so that assistance could often be too little and granted too late. The objectives of the resulting RAS 1988 were to improve the efficiency of Australian rural industry and to enhance its international competitiveness and its potential to contribute to the national economy. The RAS 1988 was not aimed at keeping farmers on the land, or propping up farm enterprises that were not viable. There were new programmes to encourage skills acquisition, adoption of new technology and enterprise switching. Re-establishment grants increased to a maximum of AUS$34,635, and household support payments, set to match the unemployment benefit in 1985, were restricted to 2 years. The household support payments were intended to be offered as a grant for the first 6 months, with payments received after that deducted from any subsequent re-establishment grant.

The administering state authorities had the discretion to convert the loan to a grant and, in some states, this was done almost as a matter of course. The increased rehabilitation grants did lead to an increase in applications and approvals, but even in 1991, with approvals at an historic high, only about 0.3% of Australian farmers obtained re-establishment grants, and in a decade (early 1980s on), only 1.2% of all farmers were re-established. In something of a 'backslide', Debt Reconstruction with Interest Subsidy (DRIS) was introduced in 1991, largely in response to the emerging drought in parts of QLD and NSW.

Following another review of the scheme, the replacement Rural Adjustment Act 1992 (RAS 92) contained sweeping changes, with the concessional loan component finally removed, though interest subsidies were to be available under the new categories of Exceptional Circumstances and the Farm Productivity programme. Farm productivity money was to be used by farmers to increase the capacity, efficiency and sustainability of farms. Successful applicants had to be viable and have formal property plans. Other programmes to boost productivity, training grants of up to AUS$500 and professional advice subsidies were to increase the managerial and technical skills of producers. There was provision for land trading, whereby state authorities could buy and sell land to speed up the process of amalgamation, or even to retire land from agricultural production. This was rarely used, except in WA.

Re-establishment grants were retained with a maximum of AUS$45,000 and there were also grants for professional relocation advice. Successful re-establishment applicants had increased from about 50 in 1984 to more than 400 in 1994, though this was still only about 0.3% of all farmers at that time. On the administrative front, the Commonwealth increased its control and established a supervisory body. In 1994, of the eight members on the Rural Adjustment Scheme Advisory Council, only two were farmers, with one representing the NFF and the other also working as a rural counsellor. The scheme was intended to promote better financial, technical and management performance from the farm sector; to provide support to farmers who had prospects of sustainable long-term profitability, with a view to improving the productivity of farm units; and to provide that support in a way that ensured that the farmers who were supported became financially independent within a reasonable period. As such, the recipients of aid effectively were being put on notice.

With the 1992 changes, household support was hived-off into a stand-alone Farm Household Support (FHS) scheme to be administered by the Department of Social Security on an agency basis. The FHS scheme provided income support at the level of the unemployment benefit for up to 2 years. The first 9 months of payments were on the basis of a grant, with subsequent payments repayable to the Commonwealth at commercial rates of interest. The loan arrangements attracted a great deal of criticism (Botterill, 2003). The Democrats moved unsuccessfully to amend the legislation in the senate to make the payment a grant rather than a loan (Cockfield and Botterill, 2006).

In 1994, the FHS legislation was amended to provide for the introduction of a drought relief payment. This was a welfare payment for farmers experiencing extreme, or 'exceptional circumstances', drought conditions. This payment was not tied to either farm viability or industry exit, representing a departure from the structural adjustment philosophy.

Under RAS 92, the federal government, in conjunction with some state governments, also introduced Regional Adjustment Strategies. The Rural Adjustment Scheme Advisory Council had identified up to 23 regional and industry 'black spots' in rural Australia. Landholders in such black spots could receive up to AUS$90,000 for re-establishment purposes, this being closer to the (1994) price of a house in a major centre. The idea was to promote rapid and extensive adjustment, thereby eventually reducing demands on governments from landholders in these regions (Cockfield and Botterill, 2006).

5.3.3 Agriculture – Advancing Australia

In 1996, the RAS was again reviewed. The inquiry found that 'RAS 92 is not appropriate to the adjustment needs of Australian agriculture in either today's business environment or that expected in the next century' (McColl *et al.*, 1997, p. ix), stating that 'RAS 92 has generally addressed symptoms rather than causes' (McColl *et al.*, 1997, p. 118). In September 1997, the Minister for Primary Industries and Energy, John Anderson, announced the AAA package.

During 1997–2007, AAA was the Coalition government's flagship policy package for assisting primary producers, farm families and rural industries to manage and adjust to change, and to position themselves better for the future. Successive reviews of the AAA package led to some refining to maintain its relevance. In 2007, a reshaped AAA package was announced, which was to last for 4 years (2007–2008 to 2010–2011).

The reshaped package was to continue the six existing AAA programmes. It was claimed that the new package would be more flexible and focused in providing information and advice, training and investment in capacity building and targeted support for primary producers, farm households, producer groups and rural industries. The package would help producers to match the performance of the best, while also providing assistance to those facing difficulties from changes. The six programmes under the package were:

- Advancing Agricultural Industries.
- Farm Business Management (FarmBis).
- Farm Management Deposits Scheme.
- Rural Financial Counselling.
- Farm Help.
- International Agricultural Cooperation.

The three programmes that provided general adjustment assistance for agricultural producers were Farm Business Management (FarmBis), Farm Management Deposits Scheme and Farm Help. More details can be found in *Future Harvest* by DAFF (2007).

5.3.4 Australia's Farming Future

The reshaped AAA package announced by the Coalition government in 2007 was, however, never put into practice. The change in government in late 2007 saw the package replaced in 2008 by *Australia's Farming Future*, a new package under the Labor government (Australian Government, 2008). As in the past when changes were made to rural adjustment assistance programmes, the new package kept many previous measures, though some with different names (e.g. when FarmBis and FarmHelp were discontinued in 2008, FarmReady was introduced in their place). However, the new package had a heavy focus on helping rural businesses adapt to climate variability. It was argued that Australia's primary industries faced unique challenges in a changing climate and could face a broad range of repercussions. There may be physical impacts (e.g. changing rainfall patterns), social impacts (e.g. changes to farm business structures, community demographics, health and well-being) and economic impacts (e.g. changing productivity levels and markets). Subsequently, the objective of *Australia's*

Farming Future was developed to equip primary producers to adapt and adjust to the impacts of climate variability.

Australia's Farming Future is an initiative by the federal government and is administered by DAFF. The package comprises a number of elements as follows, and more details can be found at DAFF (2011):

- The **Climate Change Research Program** provides funding for research projects and on-farm demonstration activities.
- **FarmReady** helps industry and primary producers develop skills and strategies to help them deal with the impacts of climate variability.
- The **Climate Change Adjustment Program** assists farmers in financial difficulty to manage the impacts of climate change. **Farm Business Analysis and Financial Assessments** and professional advice and training are tailored individually to help farmers adjust to climate change and to set goals and develop action plans to improve their financial circumstances. Rural financial counsellors can assist eligible farmers to take action to improve their long-term financial position.
- The **Climate Change Adjustment Program Re-establishment Grant**, which ceased on 30 June 2011.
- Transitional income support is linked to the climate change adjustment programme and provides short-term income support and advice and training opportunities to farmers in serious financial difficulty, while they adapt their farm to changing circumstances, including climate change.
- **Community Networks and Capacity Building** activities focus on increasing the leadership and representative capacity of target groups including women, young people, Indigenous Australians and people from culturally and linguistically diverse backgrounds.

Different from previous adjustment assistance programmes, this package placed less emphasis on helping farmers with financial difficulties but more on helping them adapt to changes caused by climate variability. This reflected the new thinking that managing farm business and financial risks are farmers' own responsibilities and government adjustment assistance programmes should focus more on helping farmers to deal with challenges that are generally beyond their control, such as climate change.

Along with the above programmes under *Australia's Farming Future*, there are also some other grants and assistance available to primary producers. Details can be found in 'Current grants and assistance' by DAFF (2012a). Among them are the flood assistance and drought assistance, which are available to farmers, their families, rural communities and eligible small businesses that serve farm production. Central to drought assistance are the various Exceptional Circumstances assistance measures, including, for example, exit grants, income support, interest rate relief and free personal and financial counselling. There is a process that regions have to go through to claim exceptional circumstances.

Most drought assistance measures are largely a continuation of the measures previously used. However, a 'Pilot of Drought Reform Measures' was rolled out on 1 July 2010 in parts of WA. It was conducted by the Australian Government in partnership with the WA government. The pilot tests a package of measures that have been developed in response to the national review of drought policy. These measures aim at moving from a crisis management approach to risk management; to support farmers, their families and rural communities better in preparing for future challenges rather than waiting until they are in crisis to offer assistance.

The drought reform measures in the pilot include (DAFF, 2012b):

- **Farm Planning** provides courses for farmers to develop or update a strategic plan for their farm business. The plan identifies priority activities to help improve the management and preparedness of the farm business to respond to future challenges.
- **Building Farm Businesses** provides grants to eligible farm businesses in two components:
    - **Business Adaptation Grants** – for eligible activities identified in the strategic plan that help farm businesses prepare for the impacts of drought, reduced water availability and a changing climate.
    - **Landcare Adaptation Grants** – for eligible activities identified in the strategic plan with an NRM focus and having a broader public benefit.
- **Stronger Rural Communities** offers a suite of measures that aim to assist rural communities to prepare for and manage agriculture-related hardship to build social capital and community networks and to encourage participation in social networks and community events.
- **Farm Social Support** provides assistance in building a better coordinated social support network to meet the mental health, counselling and other social needs of farming families and rural communities.
- **Farm Family Support** offers income support for farmers facing financial hardship, allowing them to meet basic household expenses.
- **Farm Exit Support** provides grants to support farmers who decide to sell their farm, including for retraining and relocation expenses. This assistance helps farmers and their families to make a fresh start.
- **Beyond Farming** puts current farmers in touch with former farmers to talk about opportunities outside of farming and to talk to someone who has been in the same position about the options for themselves and their families if selling the farm business or retiring.

Initially, the pilot was planned for a trial period of 12 months to end on 30 June 2011. On 10 May 2011, the Australian Government announced that the trial period was to be extended for a further 12 months until 30 June 2012. Also, as of 1 July 2011, the pilot covered an expanded area including the entire south-west region of WA.

The reform measures under the pilot are quite comprehensive and represent major measures that would be needed in Australia to support the farming community in decades to come. The Australian Government has not given any clear indication about what it may do with the pilot after it ends in June 2012.

Rural adjustment assistance measures since 2008 seem to have, to some extent, moved away from previous practices. This is reflected by the tendency to reduce inefficient subsidies on inputs and move away from crisis management to risk management. Welfare payments are now part of the normal social security provisions. Farmers who receive welfare payments are either in extreme financial difficulty (e.g. temporary household expense relief) or in the process of exiting farming.

Moving away from crisis management towards risk management represents a very important policy innovation. It is designed to help build stronger and more resilient farming communities that will prepare themselves for future challenges. Such a shift in assistance approaches seems also to have started at an appropriate time: 2008 was the year when wheat single desk export was abolished and following which few major marketing regulations existed in Australia. It was also at a time when many parts of Australia had been experiencing an unprecedented run of poor seasons, with consequent pressure for government assistance.

Without the need to deal with market regulation-related matters, shifting attention to assist the farming community in handling business and financial risks caused by adverse weather conditions and volatile markets was a logical move. According to conversations between the author and farmers in different states, farmers are generally receptive of the recent reform measures. It is anticipated that such measures will be embraced by the farming communities. The philosophy that assistance programmes should help farmers to improve their own risk management capabilities is likely to dominate the future design of adjustment assistance measures in rural Australia.

5.4 Changing Role of Government in Agriculture

From the early 1970s, when the Whitlam government started to reduce agricultural support, to 2008, when AWB single desk wheat exporting was abolished, extensive efforts have been made by various Australian governments to deregulate agriculture. Over a period of four decades since then, the rural sector has been largely deregulated. Today, Australian agriculture operates within a largely free market economy.

As a result of the deregulation, the government's role in agriculture has also evolved. It has changed from focusing on regulating and controlling production and marketing to the provision of services to facilitate farmers to produce most efficiently and sell most profitably. Government departments nowadays pay more attention to innovating and maintaining conducive institutions through which services are provided to help farmers improve their on-farm productivity and achieve better profitability by marketing their products smartly. Australian governments now devote policy resources chiefly to areas like biosecurity (including exotic pests and diseases), water markets, climate variability, regulation of natural resource use and protection, food safety, access to international markets and agricultural R&D.

At the same time, views about the role of government across the economy have narrowed. Government intervention is related more closely to market failure. Mechanisms to correct market failure other than direct provision of services by government are being devised and adopted more often.

Such a change in the government's role in agriculture is clearly reflected in the statements of both federal and state departments of agriculture about their role and their policy focus. The focus of departments of agriculture at both the federal and state levels is to provide services to facilitate rather than regulate or control rural production and marketing. Take the federal department of agriculture and the NSW DPI as an example. DAFF's role is to develop and implement policies and programmes that ensure Australia's agricultural, fisheries, food and forestry industries remain competitive, profitable and sustainable (DAFF, 2012c). Its major policies and programmes include:

- encourage and support sustainable natural resource use and management;
- protect the health and safety of plant and animal industries;
- enable industries to adapt to compete in a fast-changing international and economic environment;
- help improve market access and market performance for the agricultural and food sector;
- encourage and assist industries to adopt new technology and practices; and
- assist primary producers and the food industry to develop business and marketing skills, and to be financially self-reliant.

At the state level, the role of the NSW DPI is to improve the profitability, sustainability and skills of the agricultural sector, to manage biosecurity risks impacting on primary industries and the environment and to respond to emergencies and disasters (NSW DPI, 2009, p. 9).[4] The focus of the NSW DPI's policies and operations is to contribute to:

- strong economic performance of primary industries;
- appropriate access to, and wise management of, natural resources;
- safe, healthy and biosecure industries;
- a strong voice for primary industries; and
- excellence in people, innovation and service delivery.

It should be pointed out that while agricultural deregulation is meant to reduce or remove completely the rules and controls over farmers' production and marketing, it does not mean that all regulations have been abolished. On the contrary, some regulations are still required as part of the services to the rural community. In Australia today, government departments still carry out many regulatory activities, such as food safety standards, natural resource protection, OH&S and chemical use safety.

5.5 Concluding Comments

As a result of continued efforts to deregulate over four decades, today there is little government intervention or regulation in the production and marketing of agricultural products, both domestically and internationally. Agricultural industries operate in a largely free market economy. Australian agriculture has become one of the least intervened, subsidized and protected in the world, directly facing global competition. Today, governments in Australia focus on providing services to help farmers to produce efficiently on the farm and compete successfully globally.

There are some important and useful experiences to be gained from Australia's agricultural deregulation practice. These include:

- changing the paradigm that saw agriculture as a special industry and instead, treating it the same as any other industry;
- treating a farm operation as a business entity, the same as any other business in the economy;
- helping farmers during the process of deregulation to make adjustments to the changing market conditions; and
- finding that farmers are more resilient to change than many have thought.

5.5.1 Treating agriculture the same as any other industry

Traditionally, and as is still the case in many other countries, there is a strong tendency to treat agriculture as a special industry and hence argue that different policies are required. Australian agricultural economists and government leaders have been pioneering in breaking away from this paradigm well before the 1970s. Starting with the Whitlam government in the early 1970s, agriculture has been treated gradually as an industry that is the same as any other industry, and rural policy has been formulated in the context of economy-wide policy. Consequently, agricultural protection and assistance have been withdrawn gradually and farmers have to compete in the free market. Today, the focus of Australian agricultural policy no longer concerns using statutory marketing powers to reduce farmers' exposure to the uncertainties of global and domestic supply and demand shocks, which usually resulted in economic welfare transfers to the agricultural sector. Rather, policies are largely about the provision of necessary services to encourage the agricultural industry to be more productive and competitive and to play its role as an integral part of the broader economy. Examples of such services include NRM, biosecurity protection and negotiations for access to overseas markets.

5.5.2 Treating a farm operation the same as any other business

Farming in Australia is no longer treated as a special occupation. The Australian farming community has generally accepted that farming, just like any other business operation, faces all sorts of risks, and possible adverse weather is just part of the risks that they have to handle. Now, more focus is devoted to helping farmers manage risks using various risk management tools to assist in production and marketing decisions and in the management of a portfolio of farm and non-farm assets.

5.5.3 Helping farmers to adjust

There have been winners and losers from deregulating agriculture. The process can be a very painful one for farmers and their families and can create stress for rural communities. During deregulation, Australian governments have been proactive and innovative in devising various adjustment assistance measures, as discussed in Section 5.3. Indeed, there has been an unbroken sequence of modern rural adjustment programmes for about four decades since 1971 (Rural Reconstruction Scheme) (and with debt relief dating back further to the 1930s). These programmes, with occasional modification, are likely to continue because rural adjustment is an ongoing process. Such programmes have ensured that a safety net is in place to protect the welfare of those farmers and their families who are in financial stress; and in return, they help the ongoing rural adjustment process.

5.5.4 Farmers are more resilient to change

The Australian experience shows that the negative effects of agricultural deregulation have not been as severe as expected. Some producers left the affected industries for other agricultural activities, non-farming vocations or retirement. Producers who chose to remain were not passive market participants. They took steps to improve enterprise performance and their financial situation. The widespread industry contractions that some had feared did not occur. Instead, after a short period of adjustment, there was an improvement in industry performance, as evidenced by increasing output in almost all agricultural industries (see Chapter 2). From interviews by the author with farmers across different states in various industries, many farmers are happy to run their farm business without any government support, and many in fact expressed that they did not want any government support as that would only induce inefficiency.

Australia's approach in deregulating its agriculture offers valuable insights to both developed and developing countries around the globe. Although the approaches used in Australia may not be applicable directly, the general principles should interest economists, industry leaders and policy makers in other countries. Given that agricultural adjustment is an ongoing process, transitional assistance is always needed to facilitate change. In the meantime, political judgements are also required to judge the relative merits of the welfare, equity and economic efficiency issues associated with assistance measures.

Less government intervention in agriculture in both developed and developing countries has been difficult to achieve in the face of concerted vested interests. In developed countries, agriculture has often been subsidized to the cost of domestic consumers and producers in exporting countries. In developing countries, agriculture has often been taxed in an attempt to lower food prices to the non-farm sector, but this has given farmers little incentive to produce enough food for growing populations. The extent of government intervention has proved to be an ongoing source of conflict in negotiations such as the Doha Round to free trade between countries and across industries.

Perhaps other countries can learn from the Australian experience, which shows that the negative impacts of reducing government intervention in agriculture can be ameliorated by appropriate adjustment packages leading to more efficient use of resources throughout the economy and a more competitive agricultural sector. Economic theory suggests that interventions generally lead to lower efficiency in agriculture, resulting in a loss of national welfare.

Developing countries may also make efforts to remove where they can some existing controls over agriculture. In the meantime, agricultural economists and policy makers should be cautious when making recommendations to introduce new regulations. In particular, developing countries should try to avoid stepping into the trap of heavy agricultural protection and subsidy and then finding it hard to remove.

In developing nations, the farming population generally accounts for a large portion of the total population; yet they have a weak voice in society. The voice of rural people in poorer countries is often unheard or ignored. As a consequence, there is a lack of services to help those many small farmers to survive or to exit from farming. In this regard, governments of such economies should consider devising programmes and providing the necessary services that will help those numerous small farmers to survive and adjust.

Support measures need to be designed to help those non-viable farmers to quit farming. Otherwise, the consequence can be that many of them will be 'tied' to the land and, in many cases, remain poor for a long time. Without assisting them to exit from farming through an ongoing process, it will be much harder to deal with the accumulated problem at a later time. It is likely to be a better option for such governments to devise adjustment assistance programmes to help those non-viable farmers to leave farming and start elsewhere for a new and, hopefully, better life.

Notes

[1]The reference list in the book, *Rationalising Rustic Regulation*, by E. Sieper (1982), provides sources for a number of writings by some renowned Australian agricultural economists in the 1960s and 1970s about their different views concerning regulation, such as John Crawford, Fred Gruen, Stuart Harris, Alan Lloyd, John Longworth, Roger Mauldon and Ross Parish. Sieper (1982) is an important reading regarding Australia's agricultural regulation, regarded as 'a landmark in the discussion of agricultural policy in Australia' (Parish, 1982, p. vii). Those Australian agricultural economists who were influential in opposing regulated agricultural marketing, though responses on particular issues varied, include Keith Campbell, John Crawford, John Freebairn, Stuart Harris, Fred Gruen, Alan Lloyd, Ross Parish and Alistair Watson.

[2]The working group consisted of Dr Stuart Harris (Convenor), Sir John Crawford, Professor F.H. Gruen and Mr N.D. Honan. The report was entitled *The Principles of Rural Policy in Australia* and was published in Canberra by the Australian Government Publishing Service in 1974. The publication of this report was considered an event of considerable significance to Australian agriculture. Keith Campbell (1974) provides a critical but very interesting and thought-provoking review of the report. J.C. Tothill (1975) comments on the report from the point of view of tropical grassland agriculture.

[3]Sieper (1982) provides a definition of many of them (as used in the context of Australian agriculture) and explains how they are used and what their implications are. Especially, he tried to understand agricultural interventions in a distributional perspective.

[4]Between 2009 and 2011, the NSW DPI, together with others, was merged into a new mega department, the NSW Department of Industry and Investment, and became one division of this new department. Following the change in state government in March 2011, the DPI was separated from the mega department. The annual report of the newly independent DPI is not yet available. In passing, during the same period, the QLD DPI

was also, together with several other departments, merged into a new mega department, the QLD Department of Employment, Economic Development and Innovation (DEEDI). In March 2012, there was a change in the QLD state government. Soon after, the functions of the DEEDI were designated into several departments; in the case of agriculture, the QLD Department of Agriculture, Fisheries and Forestry. NSW and QLD were the only two states that placed agriculture into a mega department, whose name carried no words indicating agriculture or primary industries. There were concerns at that time that other states might follow suit. Given the recent reversal in both states, it is anticipated that in all states, those departments servicing agriculture will have words in their names indicating agriculture or primary industries, at least into the near future.

References

ABS (2002) *Year Book Australia*. Australian Government, Canberra.

Australian Government (2008) *Strengthening Rural and Regional Australia*. Australian Government, Canberra.

Botterill, L.C. (2003) *Government Responses to Farm Poverty 1989–1998: The Policy Development Process*. Rural Industries Research and Development Corporation Publication No 02/163, Canberra.

Campbell, K. (1974) Rural policy in Australia – 1974 style. *Australian Journal of Agricultural Economics* 18, 157–170.

Cockfield, G. and Botterill, L.C. (2006) Rural adjustment schemes: juggling politics, welfare and markets. *Australian Journal of Public Administration* 65, 70–82.

DAFF (2007) *Future Harvest*. Australian Government, Canberra.

DAFF (2011) Australia's farming future (http://www.daff.gov.au/climatechange/australias-farming-future, accessed 12 December 2011).

DAFF (2012a) Current grants and assistance (http://www.daff.gov.au/about/current-grants#rural, accessed 14 March 2012).

DAFF (2012b) Pilot of drought reform measures in Western Australia (http://www.daff.gov.au/agriculture-food/drought/drought-pilot, accessed 14 March 2012).

DAFF (2012c) About DAFF (http://www.daff.gov.au/about, accessed 1 February 2012).

Harris, D. (2005a) Industry adjustment to policy reform: a case study on the Australian dairy industry. Report prepared for the Rural Industries Research and Development Corporation, RIRDC Publication No 05/110, Canberra.

Harris, D. (2005b) Rural industry adjustment to trade-related policy reform. Report prepared for the Rural Industries Research and Development Corporation, RIRDC Publication No 05/173, Canberra.

Lewis, J.N. (1967) Agricultural price policies. In: Williams, D.B. (ed.) *Agriculture in the Australian Economy*. Sydney University Press, Sydney, Australia, pp. 299–314.

Lloyd, A.G. (1982) Agricultural price policy. In: Williams, D.B. (ed.) *Agriculture in the Australian Economy*, 2nd edn. Sydney University Press, Sydney, Australia, pp. 353–382.

McColl, J., Donald, C. and Shearer, C. (1997) *Rural Adjustment – Managing Change*. Department of Primary Industries and Energy, Canberra.

Mauldon, R.G. (1990) Price policy. In: Williams, D.B. (ed.) *Agriculture in the Australian Economy*, 3rd edn. Sydney University Press, Sydney, Australia, pp. 310–328.

Mauldon, R.G. (2009) The Industry Commission's 1991–92 inquiry into the Australian sugar industry. Presentation to the APEC training programme in resolving policy conflict, food security, structural reforms and food price inflation, 17–24 June 2009, Melbourne, Australia.

NSW DPI (2009) *NSW DPI Annual Report 2008–09*. NSW Department of Primary Industries, NSW Government, Sydney, Australia.

Parish, R. (1982) Preface to the book of Sieper, E. (1982) *Rationalising Rustic Regulation*. Centre for Independent Studies, Sydney, Australia.

Piggott, R. (1990) Agricultural marketing. In: Williams, D.B. (ed.) *Agriculture in the Australian Economy*, 3rd edn. Sydney University Press, Sydney, Australia, pp. 287–309.

PC [Productivity Commission] (1998) *Pig and Pigmeat Industries: Safeguard Action Against Imports*. Inquiry Report No 3, Canberra.

PC (2002) *Citrus Growing and Processing*. Inquiry Report No 20, Canberra.

PC (2010), *Wheat Export Marketing Arrangements*. Inquiry Report No 51, Canberra.

Sieper, E. (1982) *Rationalising Rustic Regulation*. Centre for Independent Studies, Sydney, Australia.

Tothill, J.C. (1975) Tropical grassland agriculture and the Green Paper on rural policy in Australia – a resume. *Tropical Grassland* 9, 1–8.

WEA [Wheat Exports Australia] (2011) *Report for Growers 2010/11*. Australian Government, Canberra.

6 Encouraging Entrepreneurial Farming

With the removal of regulations and controls, Australian agriculture is now truly operating in a free market. This requires farmers to be business operators just like those in any other industry. This chapter explains how the Australian governments have tried to encourage farmers to be entrepreneurs and how Australian farmers have responded.

6.1 The Need to be Entrepreneurial

When under heavy regulations and controls, Australian farmers needed just to produce and there was no need for them to look beyond their farm gate. Someone else would sell their products for them, or it was guaranteed that there would be an outlet for them to sell their products at a given price. This has changed. They now operate in a free market. Not only do they have to produce but also they have to produce what the market demands.

Beyond production and marketing, there are many other challenges farmers have to handle. One of the many challenges is the management of their financial risks. Their financial management is very complicated, due to the fact that they need to deal with financial uncertainties caused not only by production fluctuations (mainly caused by changes in weather conditions) and domestic market changes, but also by international market volatility and exchange rate changes. The other major challenge is for them to improve their productivity consistently; otherwise, their survival will come into question.

Hence, farmers in Australia have to be very entrepreneurial.

6.2 Transforming the Farming Community: From Producers to Business Operators

During the process of agricultural deregulation, the Australian governments, together with peak industry organizations and professional bodies, innovated various institutions and measures to help farmers transform themselves from being just producers to becoming business operators and entrepreneurs. This section highlights the major institutions and measures that assist farmers in becoming business operators.

6.2.1 Promoting cultural change

As agricultural deregulation progressed, a new culture was promoted gradually among the farming community. That is, the farming business is no different to any other business operation. Just like any business in many other industries, farming faces many risks and uncertainties, including adverse weather conditions and unpredictable markets. If one chooses to farm, then it is the farmer's responsibility to manage production and financial risks. Farmers need to make various changes to their practices, such as from 'Consumers buy what I want to produce' to 'I produce what the consumers will buy'.

Farmers have to look beyond their farm gates and watch closely the developments of the market for their products. In addition to being good at producing and marketing, they also need skills in managing their financial matters, collecting and analysing market and financial information, and making sound decisions for improved profitability for their operations. Such a cultural change, like other changes, initially met with resistance. Over time, farmers realized that they had no other choice. Today, passively or actively, farmers generally have accepted this cultural change. The interviews the author had with farmers in different states suggest that they are very clear about the diverse managerial skills they need to have.

6.2.2 Providing assistance

The Australian governments did not just throw farmers into the deep end and leave them to struggle for their survival. Instead, as shown in Chapter 5, various assistance measures were devised to help farmers to become entrepreneurial. This change in the nature of services has been reflected clearly in the changes in assistance programmes, as discussed in the previous chapter.

Currently, at the federal level, several programmes are available that assist farmers in becoming more entrepreneurial. These include:

- **FarmReady** under *Australia's Farming Future*. This programme helps farmers develop skills and strategies to deal with the impacts of climate variability.
- **Farm Planning** and **Business Adaptation Grants** under the **Pilot of Drought Reform Measures. Farm Planning** provides courses to help farmers to develop or update their business strategic plan. The plan identifies

priority activities to help farmers improve their management and preparedness of the farm business to respond to future challenges. For eligible activities identified in the strategic plan, **Business Adaptation Grants** may be provided to farmers to help prepare for the impacts of drought, reduced water availability and a changing climate.

- The **Export Market Development Grants** (EMDG) scheme that is administered by Austrade (Austrade, 2012). The EMDG scheme encourages small and medium-sized Australian businesses to develop export markets. It reimburses up to 50% of eligible export promotion expenses above AUS$10,000, provided that the total expenses are at least AUS$20,000. The scheme can provide up to seven grants to each eligible applicant.

At the state level, various state government departments also assist farmers with their entrepreneurial endeavours. Generally, each state government has departments or sections in agriculture, state and regional development and trade. All of them provide assistance to help farmers to be entrepreneurial. In such departments, the staff who are responsible for helping farmers to develop businesses also collaborate regularly. For example, when a farmer approaches the Department of Agriculture for assistance, the case then is often shared among the staff and there will be discussions to see how the farmer can best be helped with the information and resources available from different departments.

6.2.3 Making information available

Farmers' first major task is to produce. This task alone takes considerable time, leaving them limited time to collect and digest information for various other activities for their farming success. In Australia, farmers are provided with abundant relevant information from different sources via diverse channels, either free or for a fee.

Information sources include departments of agriculture, other government departments that provide services to the farming community, private consultants, rural suppliers and trading companies. Information is provided to farmers through diverse channels such as: rural programmes on TV, radio, the Internet, e-mail, fax, smartphone, rural show days, workshops, seminars, conferences and trade forums. Generally, for information obtained from private operators, a service fee may be involved, unless the information service is bundled with the purchase of other products.

The information that is readily available to farmers has helped the Australian farming community greatly, saving them time while enabling them to make opportune and informed decisions.

6.2.4 Rewarding entrepreneurial behaviours

These entrepreneurial farmers are well recognized and rewarded. There are various mechanisms to reward entrepreneurial farmers. Federal and state governments, peak industry organizations and professional bodies all have various programmes through which entrepreneurial farmers are nominated and rewarded. Some examples are given below.

By the governments:
- The Australian Year of the Farmer, 2012.

By industry organizations:
- The Rural Women's Award, RIRDC.
- The Australian Farmer of the Year, Kondinin Group and ABC Rural.
- Red Meat Industry Emerging Leader, AgForce Queensland.
- Rising Beef Industry Champion, Cattle Council of Australia.

By professional bodies:
- Rural Manager of the Year award, the Australian Institute of Management.

Not only do such mechanisms recognize entrepreneurial farmers, they also help other farmers to learn from such examples, encouraging more entrepreneurial behaviour.

6.3 Entrepreneurial Farmers: Success Stories

Three cases are presented in this section. They show how entrepreneurial some Australian farmers are in managing their farm business for success.

6.3.1 Moving into a new industry to start again

In 1999, Garry and Anne Fean moved into winemaking after some 20 years of cotton farming. They started cotton farming in partnership with other family members in 1982 in Moree, northern NSW. Changes in the business environment for cotton production led them to winemaking, a completely new venture for them.

A number of reasons were behind the Feans' decision to shift from cotton production to winemaking:

- The risks involved in cotton farming in Australia have become higher and are still increasing. There can be good money in cotton production. However, the investment is significant while the return is very uncertain, due to droughts, floods and also the increased resistance of insects to chemicals.
- The price of land has become very high. They could get a good price for their farms and it was a good time to sell.
- Their cotton farming was in a family partnership. The partnership will end sometime anyway.
- Labour shortage was another important reason. The booming mining industry drew many farm labourers away. They constantly had to train new labour, only to lose it.
- Water supply has become increasingly uncertain and insufficient. Cotton production is very water dependent. However, water supply increasingly became a problem for cotton production.

- The negative perception of cotton production is also a major consideration. Cotton production has been regarded by some environmentalists as a major water user and a source of environmental pollution. Hence, in today's environment, while farmers have to combat adverse weather conditions and deal with market risks, they also have added pressure from environmentalists. This has taken a lot of fun out of cotton production and made it less enjoyable.

Taking all these factors into consideration, leaving cotton farming became an attractive option to the Feans. However, it was never an easy decision to make. It took them quite a while to consider the sale of the farms in Moree. They had been in the cotton industry for some 20 years. They were not sure whether it would be the right decision to move away from cotton production and to step into a new industry with which they were unfamiliar.

The Feans' move into winemaking was due partly to Garry's early passion in winemaking, rekindled by his brother-in-law, who purchased a farm in Mudgee, NSW, in the early 1990s and developed a vineyard. In 1999, the Feans bought a small farm in the Central West Tablelands of NSW. They planted Chardonnay, Cabernet Sauvignon and Merlot vines. In 2002, they harvested their first vintage and made their first lot of wines (Cabernet Sauvignon and Merlot; the Chardonnay vintage was damaged by frost). They gave their wines the name, Angel's Garden. In the same year, they sold their cotton farms in Moree and started to focus fully on winemaking. Since then, they have increased tonnage each year and have expanded the Angel's Garden range to include Pinot Noir in 2003, fresh whites Riesling, Sauvignon Blanc and Shiraz in 2005 and Pinot Grigio in 2007.

Garry and Anne Fean made a successful transition into an industry that was completely new to them. After only 10 years, they have established a premium restaurant market for their wines. Their wines have already been regarded as of premium quality. They can sell all the wines that they can produce. In 2010, their wines were selected by Qantas Business Class Cabin. The Qantas Wine Club was also negotiating with them as an exclusive agent to sell their Chardonnay wine. As a new player in this highly competitive industry, they were indeed very successful.

Garry was asked by the author to disclose how they realized these outstanding achievements and to share the experiences that led to their success. He believes the following are important in their venture:

- **Prepared to learn.** As newcomers, they had to learn many new things by reading a large amount of literature. In 2001, Garry enrolled himself in the Bachelor of Wine Science programme at Charles Sturt University to help him learn more about the wine industry. Their first vintage from the vines they planted in 1999 would be ready for making wines in 2002. In an effort to learn and prepare themselves for the 2002 harvest, they purchased 1 t of Cabernet Sauvignon grapes in 2000 and a couple of small stainless steel tanks and barrels to experiment with winemaking. Bolstered by the experience, they purchased 5 t of Cabernet Sauvignon grapes in 2001 to strengthen their skills in winemaking. These efforts enabled them to master quickly some essential techniques to produce wine.

- **Quality control.** They produce for quality not for quantity. They regard quality control as most important. They exercise good practice to ensure they do not comprise their wine quality.
- **Set a target.** They decided to target top-end restaurants.
- **Personal approach.** There are many good wines produced in Australia. If they had sold their wines through agents in the normal wine outlets, it would be very hard for consumers to recognize their new addition to the wine market. They chose to market their wine directly to restaurants. This gave them a competitive edge over many other competitors in that they had direct personal contact with their customers. They had the opportunity to listen to the comments their customers made about their wines. This also gave them the opportunity to tell stories about their wines to restaurant staff, who could then relay these stories to their restaurant patrons. Such stories have proven to be quite effective in attracting consumers' interest in their wines.

Wine production requires a large initial investment. By 2010, they had managed to break even. Since 2011, they have been able to make a profit.

6.3.2 Diversifying businesses to mitigate risks

Riverside Sanctuary is located in the Northampton Shire in WA, about 120 km north of Geraldton. This farming property has been the home to five generations of the Porter family and is currently managed by Fleur Grieve, the fourth generation of the family. The Porter family settled a small landholding in this area in 1911, and in 1927 acquired more land where the Murchison River runs. This family farm has become known as 'Riverside', now 'Riverside Sanctuary'. The farm is 13,000 ha in area, of which 8000 ha are arable, and is fully owned by the Porter family.

For many years 'Riverside' operated as a traditional wheat and sheep farm. Sheep were allowed to graze along the portion of the Murchison River on their farm, but they were hard to manage. In the 1960s, as a management option, sheep were gradually removed from the river system. Some 20 years later, the natural recovery of these areas was vivid. Bob Porter, father of Fleur, began to realize that the farm was not suitable for sheep. The Porters started to reduce sheep numbers on their farm. By 1998, all livestock had been phased out and the farm became entirely a cropping enterprise.

Riverside Sanctuary is close to the northern border of the WA far northern wheat belt. Rainfall in this area is erratic and unreliable. This is detrimental to wheat production, which is not irrigated but is weather dependent. Table 6.1 shows the variability in rainfall on the farm, which in turn results in output fluctuations.

Clearly, relying only on wheat production to support the family financially was very risky. As can be seen from Table 6.1, in severe drought years such as 2006 and 2007, the income would be minimal. To mitigate such risks, the Porter family adopted the share-farming approach and ventured to run a farm tour business.

Share-farming helps the Porters to reduce the risks of crop production significantly. The Porter family provides the land for the share-farmer to grow crops, normally wheat and lupins. The share-farmer covers all input costs for

Table 6.1. Rainfall and wheat output variability at 'Riverside Sanctuary', WA (provided by Fleur Grieve).

Year	Area planted (ha)	Wheat harvests (t)	Wheat yield (kg/ha)	Rainfall (May–September) (mm)	Annual rainfall (mm)
2006	8,000	500	63	55	172
2007	6,000	1,100	183	101	198
2008	6,000	10,000	1,667	158	325
2009	6,500	7,200	1,108	202	230

crop production, including wages for hiring farmhands, seeds, chemicals, etc. From the harvests, the Porters receive 10% of the lupins and 20% of the wheat. The rest belongs to the share-farmer. Each party is responsible for selling its own share of the products.

What the Porters get is clearly dependent on the harvests. In a poor year, the income from crops can be very minimal. However, the Porters' outgoings are zero, helping them to avoid financial losses in a very poor year. For example, in 2006 and 2007, the harvests could hardly cover the costs of the inputs. So far, both parties are happy with the share-farming arrangement and have a good relationship between them. In fact, the share-farmer is invited to participate in the farm's long-term planning to ensure the viability of future crop production on the farm.

Tour operation provides the family with additional non-farming income. In 2000, the Porters decided to embark on a farmstay tourism venture. Existing houses on the farm were renovated for guest accommodation. Various tour activities were gradually developed, including the Shearing Shed BBQ area and walk trails in the protected section of the Murchison River. In May 2001, the first group of guests stayed and the business grew from there.

The move into tourism has meant the Porters are dealing with more people, sometimes in large groups, who require various kinds of attention. This presented some challenges to the Porters. Fortunately, the warm Australian country hospitality with which they have all grown up has helped them to win over many happy visitors. Today, Riverside Sanctuary is an accredited tourism business. It has won several awards for its excellent services to visitors, including: 'Midwest Tourism Awards 2002'; Winner B&B Farmstays, 'Banksia Awards 2004'; Winner Environmental Leadership and 'Take A Break Awards 2007'; and Winner Best Places to Stay. It was also a finalist in the Ecotourism Section of the 'WA Tourism Awards 2004'.

The Riverside Sanctuary experience focuses on providing opportunities for visitors to experience the wonders of nature, to learn ways of managing the human impact upon it and to relax in a beautiful place in the Midwest of WA. As part of the tour's objectives, the environment sustainability concept is promoted to visitors. As a demonstration, at Riverside Sanctuary, all electricity used is 100% green energy sourced through Synergy, the local electricity provider; all the water for washing, showers and toilets is pumped from underground to the homestead by windmill; all drinking water is natural rainwater. And, for every guest who stays with the Sanctuary, the Sanctuary plants one tree – a long-lived, carbon storing, drought resistant Australian eucalypt.

In addition to income from crop production and tour operation, the family also receives income from off-farm investments in shares and properties. Financially, the Porters are debt free and secure. With diverse sources of income, the family is always 'safe', even when the most severe adverse weather hits.

6.3.3 Integrating vertically to market organic beef

The Jonssons' venture beyond their farm gate into food processing and marketing provides another story that demonstrates the desire of many Australian farmers to be entrepreneurial.

Kerry and Greg Jonsson, with six grown-up children, own Jervoise Station, which is situated nearby Greenvale in north QLD, approximately 300 km from Townsville. The station is on the waters of the Burdekin, just on the eastern side of the Great Dividing Range and due west of Ingham. The property is 70,000 acres in area and is suitable for the production of a range of crops and animals. In the late 1970s, Greg realized that the use of chemicals in the raising of cattle was not good for anyone except the chemical manufacturers. He decided not to use any chemicals at all in his production of cattle. His effort led Jervoise to being awarded in 1996 a certified organic beef station by the Biological Farmers of Australia.

The certification is an official recognition that their herd is totally free of any dips, fly sprays, drenches, hormonal growth promotants or injections of any type. However, this recognition did not bring any immediate financial gains to the station. This was due simply to the fact that there were no certified organic abattoirs in Far North QLD. Organically produced beef that is processed through non-certified organic abattoirs cannot be claimed to be organic. In 2004, the Jonssons bought a small abattoir in Tully, about 250 km north of Townsville, and added meat processing facilities to it. In mid-2005, their expanded facilities in Tully began to operate. Soon, the abattoir was certified as an organic abattoir. The abattoir is still the only certified organic abattoir in Far North QLD.

With the meat processing plant at Tully and both the plant and the station being fully certified for organic meat production, Jervoise is in a position to boast about its 100% Australian organic beef. They now supply their organic beef products to consumers in the Far North QLD region through outlets in Townsville, Cairns, Mareeba and Atherton. To increase the awareness of the public about their organically produced beef, they also have stands at the farmers' markets in Townsville and Cairns, where they interact directly with potential customers.

The Jonssons are also very innovative in providing convenience for customers to buy and consume their products. Customers are welcome to discuss with Jervoise their meat purchase in terms of price and delivery for a full range of organic beef products. Jervoise has the capacity to supply everything from full container loads to one carton for a restaurant anywhere in QLD, or for individual orders for many retail customers.

To encourage customers to call them to discuss any issues or to place orders, they offer a free call, which is based on VoIP facilities that are available on the Internet. Customers simply just click on the ClickCalling link on Jervoise's website, enter their telephone number and they will then be connected to the station immediately by telephone. No call charges apply to the customer for this call from any fixed or mobile phones.

Stepping out of one's comfortable turf often is a huge challenge. At an interview with the key players – Kerry, Greg and Kristine – one of the daughters of Kerry and Greg indicated that, at the very beginning, they were excited but also very nervous, simply because cattle slaughtering and processing and retailing was completely new territory to them. They were not so confident about the outcome. However, they allowed for errors and were prepared to learn from those errors.

For the Jonssons, the biggest challenges when they first moved beyond their farm gate included the following:

- **Lack of knowledge.** Kerry and Kristine were to manage the abattoir and processing but they were graziers and had no idea about meat processing. They worked hard and obtained their meat inspector licence. Kristine also undertook courses to qualify as a certified butcher.
- **Financial tightness.** Like many other farmers in Australia, they were asset rich but cash poor. To start the vertical integration, they were on a shoestring budget but they worked very hard and did many things themselves to reduce costs.
- **Being women.** Meat processing is a heavily male-dominated industry. Kristine and Kerry had to cope with such dominance and demonstrate their capability.

The Jonssons believe that they have been successful in their venture and financially it has been getting better and better. When asked why they work so hard, the answer is that they of course want to earn money from their work but they also obtain personal satisfaction from providing 100% organic and chemical-free beef products to their customers.

Looking into the future, they are confident and optimistic. More and more people are getting to know their products and appreciate the benefits of organic foods. Since 2005, their business has been growing steadily. In 2012, their business has grown much faster because they have expanded the products they sell. Now they sell not only their own organic beef products but also organic chicken, lamb, goat meat and pork, which they source from other certified organic producers. In the near future, they may extend to produce organic goats by themselves on their own station at Greenvale.

Their meat processing plant has a capacity for 40 head of cattle each week. Currently, they process only four head of cattle, which is what they can sell in the Cairns and Townsville regions. The Jonssons plan to supply around 20 head of prime certified cattle a week to markets in the south, 50 weeks a year, while still servicing the Cairns and Townsville regions.

6.4 Concluding Comments

Prior to the 1980s, few farmers in Australia needed to be entrepreneurial. Those farmers in heavily regulated and controlled industries were well insulated from the market. Since the early 1980s, as deregulation and the removal of controls progressed, being entrepreneurial gradually became more necessary for farmers to survive and to survive well. The Australian governments and agribusiness firms provided some assistance to farmers in becoming entrepreneurial. The few examples presented in this chapter show that Australian farmers are capable of being entrepreneurial and they are rewarded for being entrepreneurial through personal satisfaction, improved income and public recognition.

In Australia, fostering farmers' entrepreneurship has played an important role in helping them to survive without any government assistance. Farmers' resilience and ability to survive in tough markets must not be underestimated. Reducing and even removing government intervention and protection in agriculture is achievable in both developed and developing countries. What is important is to create an environment in which farmers are encouraged and facilitated to be entrepreneurial.

In general, farmers everywhere are intrinsically entrepreneurial. However, compared to their counterparts in developing countries, farmers in developed economies have better access to information and wider exposure to markets. On the other hand, greater efforts may be needed to help farmers in developing countries to adapt when interventions and controls are removed. None the less, fostering farmers to become entrepreneurial in preparation for the removal of controls and interventions will be most beneficial in both developing and developed countries.

Reference

Austrade (2012) What is EMDG? Australian Trade Commission (http://www.austrade.gov.au/What-Is-EMDG/default.aspx, accessed 14 May 2012).

7 Proactive R&D Investments

Agricultural R&D is another important factor that has contributed to Australia's agricultural success. R&D has promoted productivity growth significantly in Australian agriculture over the past 50 years (Nossal and Sheng, 2010; Sheng *et al.*, 2011). This is attributable partly to the country's proactive investment in agricultural R&D. Today, diverse arrangements exist for agricultural R&D. One of the most innovative arrangements is the creation of rural R&D corporations, which has proven to be most successful. This chapter shows how Australia invests in agricultural R&D, contributing to its agricultural productivity growth.

7.1 Agricultural Research, Development and Extension in Australia: An Overview

Addressing agricultural R&D issues without including agricultural extension is incomplete. Elsewhere in this book, the term 'agricultural R&D' is used for simplicity in discussion; however, whenever 'R&D' is referred to, unless otherwise indicated, it includes 'extension'.

Australia has invested in agricultural RD&E through various mechanisms. The planning, funding and delivering of agricultural RD&E in Australia is highly complex. There are multiple funders and suppliers. This section provides an overview of the broad framework of RD&E and the major players in it: funders, administrators, and procurers and suppliers. Figure 7.1 highlights the RD&E funding and delivery arrangements in Australia.

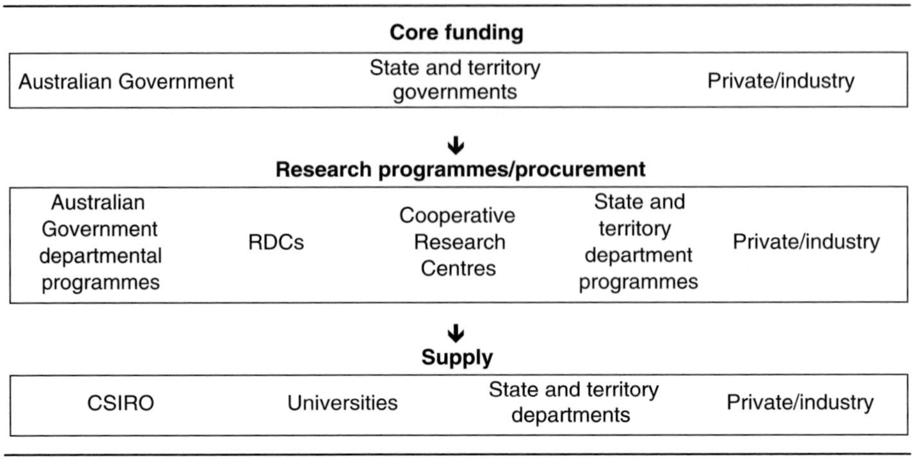

Fig. 7.1. Agricultural R&D funding and delivery framework (PC, 2011, p. 11).

7.1.1 Sources of funding

Australia's PC recently estimated that in 2008–2009, total funding for rural R&D (including farming, forestry, fisheries and some processing) was in the order of AUS$1.5bn (equivalent to about 3.3% of rural gross value of production (GVP)) (PC, 2011, p. 12). Seventy-six percent of the funding was from government sources (federal government: 48%; state and territory governments: 28%) (see Table 7.1). The remaining 24% was from private sources.

7.1.2 Rural R&D programmes and procurement

Federal government

The Australian Government provides funds for rural R&D through a range of programmes spread across several departments. The largest of these programmes is the Research and Development Corporation (RDC) programme, which is administered by DAFF. The RDC programme is a co-investment model whereby the RDCs collect industry levies which are matched by the government up to a cap of 0.5% of GVP. The RDC's model is a unique Australian innovation and has been well regarded, both in Australia and internationally. Section 7.2 will provide further explanations about how this model promotes and facilitates rural R&D in Australia.

The Commonwealth Scientific and Industrial Research Organization (CSIRO) is provided by the federal government with the second largest block funding for its rural-related R&D activities. CSIRO is Australia's national science agency. It was formed in 1926 and has since played a significant role in contributing to Australia's agricultural R&D. It is the largest supplier of rural R&D in Australia.

Table 7.1. Rural R&D funding in Australia, 2008–2009[a] (PC, 2011, p. 13).

Organization type	Funding (AUS$ million)	Share (%)
Australian Government[b]		
Cooperative Research Centres	63	
Core funding for CSIRO	193	
Core funding for universities[c]	118	
Research and Development Corporations (RDCs)	218	
Other departmental programmes[d]	114	
Foregone tax receipts arising from R&D tax concessions	9	
Total Australian Government	715	48
State and territory governments		
Project-related budget allocations[e]	348	
Capital investment in R&D facilities	47	
Payments to other funders and suppliers	21	
Total state and territory governments	416	28
Private/industry		
Levy payments provided to RDCs	248	
Other (for which a tax concession is claimed)[f]	116	
Total private/industry	364	24
Total	1495	100

Notes: [a]This does not include funding from royalties and other intellectual property income (on the basis that these have been generated by past funding from governments and private parties). Also, the data do not include in-kind contributions from the private sector, such as through the provision of land and facilities for experiments. [b]Only the portion of the budget assigned to rural R&D is included. [c]Estimated by applying the rural share of total university funding received from contestable sources and the portion of university students studying in agriculture-related areas to the three largest university block grants. [d]Includes programmes aimed at wider issues (such as climate change), programmes with no sector-specific focus and any one-off payments. [e]Includes rural R&D and associated extension funding for programmes facilitated within the primary industry department (or its equivalent). Any funding for rural R&D from state and territory government environment departments and the like is not included. [f]Calculated using tax concession data (including an estimate for concessions claimed for R&D on agricultural chemicals). Also includes payments made to the Australian Animal Health Laboratory.

Universities and the Cooperative Research Centres (CRCs) are two other major recipients of funding from the federal government for rural R&D. Universities in Australia today play an increasingly important role in rural R&D. Many rural R&D activities by universities are carried out in collaboration with other research providers. CRCs are partnerships between different research funders, suppliers and end users, formed to develop and undertake R&D in specific areas, with a particular emphasis on applied R&D. CRCs must include a university and an end user, with an RDC, CSIRO, industry representative or government organization being among the other possible partners. CRCs receive public funding, which must be matched by participants' cash and in-kind contributions, for a period of up to 10 years via a competitive merit-based selection process. The CRC programmes are administered by the Department of Innovation, Industry, Science and Research. Further details about the CRCs will be provided in Section 7.3.

The Australian Government also provides support for rural R&D through a range of other programmes. While some of these programmes are industry specific (such as the Fisheries Resources Research Fund), most are general. Some of these programmes target issues of direct relevance to the wider rural sector (for example, the Climate Change Research Program). In other cases, there is no sector-specific focus, but the rural sector may none the less receive some funding support (for instance, the Commercial Ready program and the R&D tax concession). Table 7.2 lists major Australian Government programmes through which funding for rural R&D is provided.

State and territory governments

The state and territory governments provide funding for rural R&D, chiefly for in-house research conducted in state and territory research institutes and experiment stations and related extension activities. In addition, they also contribute some funding (or in-kind contributions) to the CRCs and RDCs.

Private parties

Private funding for rural R&D is from various sources. The share of private funding varies considerably across industries. Most of the research sponsored by private entities is at the more applied end of the R&D spectrum (Keogh, 2011).

7.1.3 Providers of rural R&D in Australia

There are five major players in Australia that supply rural R&D and extension: the Australian Government, state and territory governments, CSIRO, universities and private providers.

The Australian Government

Provision of R&D is not a major function of the Australian Government. However, several federal departments carry out activities that are extension and educational in nature through their participation and displays in agricultural field days, rural shows, farmer conferences, etc. One major rural extension and education service provided by the federal government, worth particular mention, is the ABC radio rural programmes and the Landline programme on ABC television. The only federal department that has an agency that provides serious rural-related research is DAFF. Under its jurisdiction is the Australian Bureau of Agricultural and Resource Economics (ABARE). ABARE is involved in research directed at the identification and evaluation of policy options, with the main purpose of providing information and research inputs to facilitate agricultural policy decisions. ABARE also researches economic issues of direct relevance to primary producers. It has been an outstanding economic research agency in Australia. Previously,

Table 7.2. Australian Government programmes providing funding for rural R&D[a] (PC, 2011, p. 18).

Portfolio	Programme
Agriculture, Fisheries and Forestry[b]	Australian Pest Animal Research
	Caring for our Country
	Climate Change Research
	Fisheries Resources Research Fund
	Forest Industries Climate Change Research Fund
	Regional Food Producers Innovation and Productivity
	Research and Development Corporations
Climate Change and Energy Efficiency	Australian Climate Change Science
	Bilateral Climate Change Partnerships
	Greenhouse Action in Regional Australia
Innovation, Industry, Science and Research	Australian Research Council Grants
	Climate Ready[c]
	Cooperative Research Centres
	CSIRO Block Funding
	Super Science Initiative
	National Collaborative Research Infrastructure Strategy
	North West and Northern Tasmania Innovation Fund
	R&D Tax Concession
	R&D Tax Offset
	University Block Funding[d]
Sustainability, Environment, Water, Population and Communities	National Environmental Research[e]
	Sustainable Rural Water Use and Infrastructure
	Water Resource Assessment and Research Grants

Notes: [a]The Department of Foreign Affairs and Trade also provides funding to Australian entities to perform R&D related to Australian aid programmes. Some programmes are collaborative initiatives which attract investment from sources other than the Australian Government. [b]Two other programmes that until very recently were funded through the Agriculture, Fisheries and Forestry portfolio were the Advancing Agricultural Industries programme and National Resource Innovation Grants. [c]Closed for applications. [d]The Education, Employment and Workplace Relations portfolio also funds universities via schemes that support capital development and education provision in higher education institutions. [e]Previously Commonwealth Environment Research Facilities.

the Bureau of Rural Science was also a research agency under DAFF but in July 2010 it was merged into ABARE; hence, the name change since then to ABARES (the Australian Bureau of Agricultural and Resource Economics and Sciences).

State and territory governments

At the state or territory level, rural R&D is provided mainly by state departments of agriculture. One of the major functions of such departments is to conduct R&D that is relevant to the various agricultural industries in their own states. They also collaborate on rural R&D that is of relevance across states. Provision of agricultural extension services is part of their function, but their extension service has been declining.

CSIRO

CSIRO has 13 divisions. Three of them contribute directly towards rural R&D.

- Plant Industry Division: conducts research to promote profitable and sustainable agrifood, fibre and horticultural industries, develops new plant products and improves NRM.
- Livestock Industries Division: supports Australia's livestock and allied industries to become stronger global competitors.
- Land and Water Division: researches ways to manage Australia's land and water resources better, and improve the quality of the natural and built environments.

Work by some other divisions also contributes to agricultural R&D and overall agricultural development. For example, the Ecosystem Sciences Division conducts R&D across a range of landscapes, targeting social, economic and environmental sustainability. The Food and Nutritional Sciences Division develops high-quality, healthy foods that are preferred by consumers and industry, helping agricultural industries to educate the users of their products (CSIRO, 2011). Internationally, CSIRO is ranked in the world's top institutions in several research fields related to agriculture (Kerin, 2010).

Universities

Universities have also been providers of rural research (relatively small until the advent of the CRCs). They also used to provide extension services to the rural community. Due to cuts in university funding in recent years and the lack of recognition of such services within university performance evaluation systems, extension service by university academics has dropped significantly.

Private providers

Private supply of agricultural R&D takes two broad forms: (i) some rural industries are served by industry-owned providers – Bureau of Sugar Experiment Stations Limited (BSES) and Australian Wine Research Institute are examples (they receive funds from their respective industries and provide R&D services to these industries); (ii) large farming operations and multinational chemical and fertilizer companies also conduct agricultural R&D in-house, while also procuring research from other suppliers.

Provided above is an overview of the broad framework on how agricultural R&D in Australia is funded and delivered. In the following sections, emphasis will be placed on several components within this broad framework that have distinct Australian characteristics.

7.2 Rural R&D Corporations

Currently, there are 15 rural RDCs, covering virtually all of the agricultural industries. The rural RDC model is a unique Australian innovation. Rural RDCs were established and supported by the Australian Government to provide an industry-driven, market-responsive approach to rural R&D and to provide a mechanism by which farmers could fund R&D collectively – a mechanism to ameliorate market failure in the provision of research. Their main role within Australia's broad rural R&D framework is to procure research from other institutions on behalf of the industries and the Australian Government. The RDCs are funded primarily by industry levies and Australian Government contributions, with the latter mainly on a matching basis up to a limit of 0.5% of industry GVP.

In early 2010, the Australian Government referred rural RDC arrangements to the PC for inquiry. The PC published its final inquiry report in February 2011 (PC, 2011). This PC report contains a great deal of detail concerning many aspects of the RDCs in Australia. It is an invaluable source for looking into the Australian RDC model. Much of the discussion in this section is based on information available in the 2011 PC report (PC, 2011) and on the website of the Council of Rural Research and Development Corporations (CRRDC, 2012).

7.2.1 Origin and evolution of the rural RDCs

One of the most important features of the RDC model is its co-investment with funding provided by industries (levies collected from producers) and by the Australian Government. Contributing to R&D investment through producer levies has a long history in Australia. The first of such levy regimes was a state-based levy for funding the BSES, which was a compulsory scheme in place between 1900 and 1997 (BSES, 2010). There were also other regimes that were funded by voluntary contributions from producers. In 1936, the Australian Government established a compulsory producer levy for funding wool promotion and research. The wool industry model evolved over 20 years into a system whereby the government matched the industry's levy contributions and a statutory advisory committee administered the funds. This model remained in place until the mid-1980s, during which time similar schemes were introduced in other rural industries.

The *Rural Industries Research Act 1985* (Commonwealth) reformed the operating environment for sponsoring industry-focused rural R&D, creating the precursor to the current RDC arrangements. Under this Act, 14 industry research councils were established. These councils allocated funds among research suppliers on behalf of specific commodity groups. They were accountable to the Australian Government for the expenditure of matching contributions. Additionally, the Act established uniform funding arrangements across most industries (PC, 2011, p. 24).

Despite these changes, concerns persisted about how rural R&D funds were being administered, including a perceived lack of coordination and communication between the various councils and lack of clarity in their decision-making processes. Additionally, the government considered that the councils needed to develop both greater links with industry and a commercial viability (PC, 2011, p. 24).

To help address these concerns, the *Primary Industries and Energy Research and Development Act 1989 (Commonwealth) (the PIERD Act)* established the current statutory model for the RDCs. This saw the replacement of the industry research councils with the RDCs, while maintaining the previous funding arrangements. (R&D corporations had already been established in the meat and horticulture industries in 1985 and 1987, respectively.) The corporation model was premised on the need to give the RDCs operating and financial flexibility and to increase the efficiency with which R&D funds were spent. More generally, the RDC model was designed to reveal industries' research priorities better, avoiding a reliance on researchers to set the agenda, as was perceived to have occurred under the previous model (Kerin, 2010).

Mr John Kerin, the then Federal Minister for Primary Industries and Energy (from 1983 to 1991), was a key driver behind the establishment of the RDCs. With his farming background and training as an agricultural economist, his passion and contribution to promote agricultural industries through increased R&D investment is regarded highly by many Australians. Though retired from federal politics for many years, Mr John Kerin is still passionately attached to, and keenly contributes to, the development of rural R&D in Australia, as reflected in a speech he gave to the University of Melbourne in May 2010, entitled 'What policy framework would I now establish for agricultural research, development and extension if I were still Minister for Agriculture, Fisheries and Forestry'. Today, the RDC model is regarded by many as a Kerin legacy (see, for example, Kefford, 2011).

Since the introduction of the PIERD Act, two cross-sectoral RDCs, the Energy RDC and Land and Water Australia (LWA), ceased operation in 1999 and 2009, respectively. In 2003, the Tobacco RDC ceased operation. On the other hand, several other RDCs emerged. Figure 7.2 shows the timeline of the RDCs by industry.

The arrangements governing the operations of the RDCs have changed over time. A number of these changes reflect the characteristics of the particular industries concerned, including agri-political factors. But the most fundamental changes have come through the transformation of many of the original statutory authorities into industry-owned corporations (IOCs), operating under the *Corporations Act 2001* (Commonwealth), resulting in variation in the legislative underpinnings of the RDCs. The impetus for the creation of IOCs – which provide services additional to R&D – came from a desire by some industries to integrate separate R&D and marketing bodies.

Today, among the 15 RDCs, nine are IOCs. All except one of the RDCs cover particular industry sectors, such as fisheries, grains and horticulture. The exception is the RIRDC, which covers a variety of diverse, generally smaller industries, as well as sponsoring research on national rural issues. The other RDCs also invest, to a varying extent, in R&D in areas with relevance beyond their immediate industry constituency.

7.2.2 RDC funding arrangements

Most of the RDCs' funds come via industry contributions and direct payments from the Australian Government. Other sources of revenue include royalties, funding from Australian and state and territory government R&D programmes (where the RDC is procuring research of relevance to those programmes) and other RDCs (where research is sponsored on a collaborative basis).

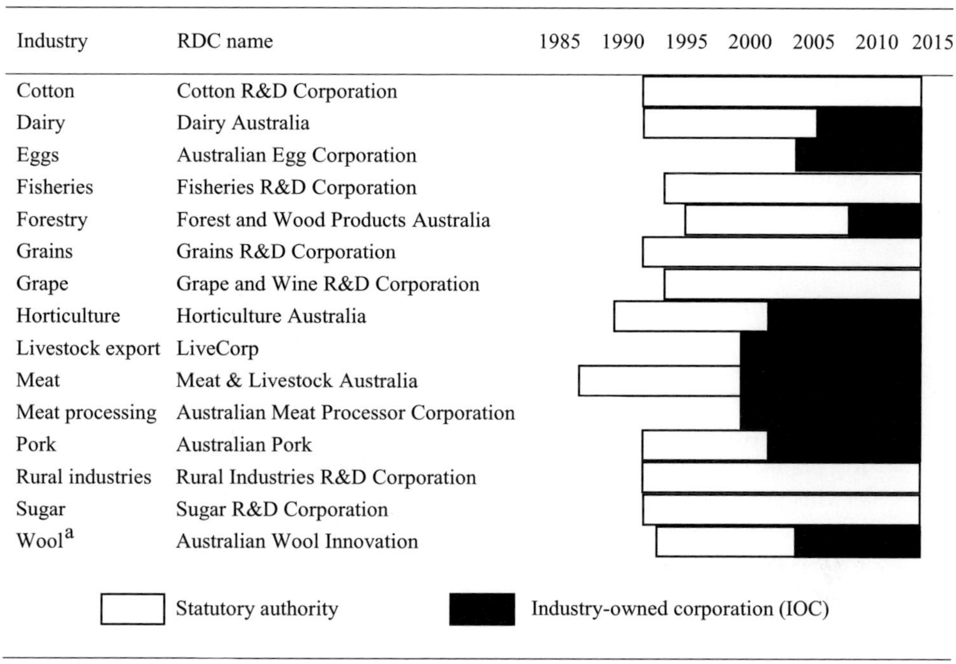

Fig. 7.2. R&D corporation timeline, by industry (based on PC, 2011, p. 25 and CRRDC, 2012). *Note*: [a]The Australian Wool Corporation operated from 1973 to 1991.

Industry contributions

The RDCs receive industry contributions from both statutory levies on producers and voluntary payments. DAFF collects statutory levies on behalf of the RDCs, charging a collection fee for this service. Voluntary contributions are collected mainly by the Fisheries RDC (FRDC), RIRDC, Horticulture Australia Limited (HAL) and Meat and Livestock Australia (MLA).

Most industries have voted to set levy rates that generate revenue close to the government's matching contribution cap (so that they can maximize the match pay from the government). However, in some industries such as grains and fisheries, as well as some smaller industries within the RIRDC umbrella, levy payments exceed the contribution cap. In the case of grains, due to higher output level, the levy collection is large. For those smaller and emerging new industries, due to smaller industry output, a slightly higher levy per unit of output may be collected for the purpose of R&D.

The Australian Government's matching contribution

In most cases, the government matches industry levies on a one-for-one basis up to 0.5% of industry GVP. This limit is calculated using a 3-year rolling average of GVP, so in practice, government contributions can exceed industry levies in any given year. The rolling average formula is used to dampen fluctuations in funding resulting from volatility in industry output levels, and hence levy payments. The matching contribution is paid on acquitted R&D expenditure, rather than levy revenue.

The IOCs receive industry levies and matching contributions via a Statutory Funding Agreement (SFA) with the government. These agreements, which differ slightly according to the particular circumstances of individual IOCs, require the entities concerned to use funds transparently and comply with various reporting and planning requirements.

7.2.3 RDC governance

Governance

Broadly, the RDC governance arrangements involve the translation of industry and government research priorities into 5-year strategic plans and annual operating plans, with after-the-event reporting on outcomes and performance. As part of this governance regime, there are various formal and informal consultation processes through which the government and industry can have input into the R&D portfolios pursued by the RDCs.

Boards

RDCs are governed by boards of directors who are generally nominated by independent selection committees. The PIERD Act requires that statutory RDC board members, including the Chairperson, be appointed by the Minister for Agriculture, Fisheries and Forestry. In contrast, IOC directors are elected by their company's members and in turn elect the Chairperson in keeping with corporations law. In both cases, the Managing Director or CEO is appointed by the board.

While there was previously a requirement that a designated 'government director' – often a public servant – sit on the board of the statutory RDCs, this requirement was removed in 2006 following the Uhrig review into the corporate governance of statutory authorities (see PC, 2011, p. 235). However, a government representative sometimes attends the board meetings of some RDCs as an observer.

Priority setting

There are various channels through which industry and the Australian Government provide input into the RDCs' priorities (Fig. 7.3). Also, the CRRDC provides an opportunity for the RDCs to collaborate on their respective strategic directions.

All RDCs must produce 5-year strategic plans detailing how industry and the Australian Government's priorities will be met and an annual operational plan specifying the general categories of R&D activities which will be funded that year, likely administrative expenses and expected receipts. While all RDCs are required to make available their strategic and operating plans to industry and the government, only the statutory authorities must have these documents formally approved by the minister responsible for agriculture.

The Australian Government's main guidance in regard to RDCs' research focus comes via the national and rural research priorities. These priorities are very broad and intentionally leave the RDCs with considerable autonomy in the

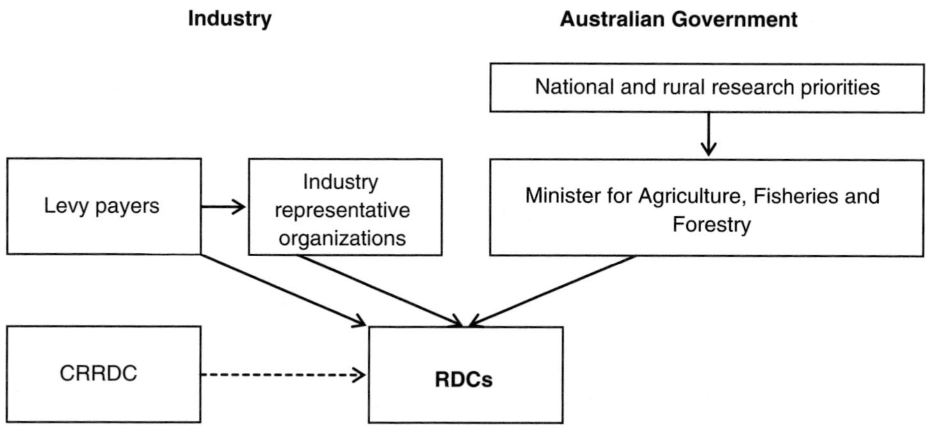

Fig. 7.3. RDC priority-setting framework (PC, 2011, p. 31).

selection of projects. DAFF has periodic meetings with the RDCs (either via the CRRDC or on an individual basis), which can provide an opportunity to clarify and reinforce the government's priorities.

The formal arrangements relating to consultation with industry in the development of 5-year plans vary between the statutory RDCs and the IOCs. The statutory RDCs must consult with nominated industry representative organizations on the development of research priorities, whereas for the IOCs there is simply a requirement in their respective SFAs to consult with industry representatives and levy payers. The statutory RDCs are not limited to consulting with the prescribed bodies only and typically engage with a wide cross section of industry interests. Also, requirements to consult with particular peak bodies are written into the constitutions of some IOCs.

The RDCs use a variety of methods to consult with industry representatives and, in some cases, directly with producers. Communication and feedback is facilitated via state conferences, newsletters and surveys. Some RDCs have established dedicated regional forums to obtain stakeholder input, and others are required to conduct regular industry polls to determine levy rates.

While the emphasis of consultation is mainly on primary producers and their representatives, some RDCs – especially those who receive levies from processors – also obtain feedback from other parts of the value chain.

Reporting and evaluation

Although all of the RDCs are subject to some general performance monitoring, these arrangements differ for the statutory corporations and IOCs.

- The PIERD Act requires the statutory authorities to provide the minister and industry representative organizations with an annual report detailing, among other things, an assessment of the extent to which their operations have contributed to the strategic and annual operational plans. These reports are tabled in Parliament. Additionally, the statutory RDCs are subject to the accountability and reporting requirements specified in the *Commonwealth Authorities and Companies Act 1997*.

- IOCs are required to report annually to the minister on their compliance with the SFA and must also have their performance reviewed periodically by independent consultants. These requirements are on top of the annual reporting obligations specified in the Corporations Act. The annual reports, compliance reports and SFAs of Dairy Australia and LiveCorp must be tabled in Parliament.

In meeting their reporting requirements, some of the RDCs (such as the Grains RDC (GRDC), RIRDC and previously LWA), have a long history of formal *ex post* project evaluation. For other RDCs, such evaluation is a more recent development under the auspices of the evaluation programme initiated by the CRRDC in 2007. This programme, and its underlying evaluation framework, seeks to quantify or otherwise indicate the impact of past RDC investments by analysing a sample of projects each year (PC, 2011, p. 245).

7.2.4 Collaboration of RDCs

RDCs collaborate widely with multiple stakeholders. DAFF (2010) believes that it is the fundamental role of the RDCs to collaborate with research providers and other funders so that research efforts are better coordinated. Collaboration between the RDCs occurs on both an informal basis and in meeting legislative requirements. Informal initiatives mostly involve engagement between RDCs on particular projects and programmes. The PIERD Act requirement that the RDCs meet at least annually to coordinate R&D activities is fulfilled by the CRRDC. While the IOCs are not required formally to attend these meetings, all are usually present. The CRRDC now has an independent chair and a full-time secretariat, and is currently performing a coordinating role in regard to matters such as evaluating and improving the administrative efficiency of RDC activities (CRRDC, 2010; PC, 2011, p. 33).

The RDCs also collaborate, to varying degrees, with:

- R&D providers seeking cash funding, such as the universities;
- partners involved in research funded through other Australian Government programmes;
- other funders of R&D, such as the Australian Centre for International Agricultural Research; and
- international rural R&D organizations. For example, Dairy Australia has Memorandums of Understanding with rural research providers in Europe and New Zealand; and MLA, GRDC and HAL have also participated in some joint funding agreements with international research entities.

Through such collaboration, and their involvement in processes such as the National Primary Industries RD&E Framework initiative, the RDCs are widely seen as having a much more significant role within the rural R&D framework than their direct funding would indicate. For example, data supplied by Cotton Australia indicate that while the Cotton RDC accounts for around 20% of total

R&D funding in the cotton industry, through funding partnerships it is involved in about 60% of all R&D projects carried out (PC, 2011, p. 34).

7.2.5 Uniqueness and advantages of the RDC model

Australia's RDC model is unique and offers a number of advantages. While other developed countries employ levies on various primary products, the organizations that are funded by such levies differ from Australia's RDCs in various ways.

- The Australian Government provides matching funds to producer levies. Few other countries provide matching public funding for levy contributions.
- The RDC model is well supported and well regarded by industry constituents and other stakeholders (Keogh, 2011). It is highly regarded internationally as well.
- Australia's RDCs have greater spending capacity. For instance, while GRDC has an annual budget of around AUS$120m, grains research organizations such as HGCA (UK) and the Western Grains Research Foundation (Canada) have budgets of around AUS$10m and AUS$5m, respectively (PC, 2011, p. 41).
- The RDCs commission agricultural R&D on a competitive basis among public and private providers. This helps ensure that the most capable providers win the bid.
- The RDC model gives rural industries, through board representation, the opportunity to have some input into research priority setting, maximizing the benefits of their investments.
- RDCs fund R&D mainly into production (on-farm) research, but also some funds are used for processing (off-farm) issues and fund portfolios of projects that have a mix of both public good and industry good components, given the taxpayer contributions.
- The RDC model encourages accountability, which in turn gives producers the confidence to continue to invest in rural R&D through paying their levies.

7.3 Collaboration and Coordination in Rural R&D

In the past two decades, collaboration has been a buzzword in the broader Australian society. Research is no exception, including rural research. Research collaboration has been promoted keenly and often funds have been made available to encourage establishing collaboration between research funders, programme administrators and providers. Rural R&D collaboration is encouraged through three major formal arrangements in Australia: RDCs, CRCs and the Australian Centre for International Agricultural Research (ACIAR).

As mentioned earlier, RDCs are engaged actively in collaboration among themselves, as well as with various R&D providers. Another major effort by the Australian Government to promote research collaboration is the establishment of many CRCs. The CRC programme is an Australian Government initiative administered by the Department of Industry, Innovation, Science, Research and Tertiary Education. The CRC programme supports end user-driven research collaborations to address major challenges facing Australia. A CRC is an incorporated or unincorporated organization, formed through collaborative partnerships between publicly funded researchers and end users. CRCs must comprise at least one Australian end user (either from the private, public or community sector) and one Australian higher education institution (or research institute affiliated with a university). In 2011–2012, there were 44 CRCs in four industry areas:

- agriculture, forestry and fishing (11);
- manufacturing (5);
- mining (4); and
- services (24).

The 11 CRCs in agriculture, forestry and fishing are:

- Australian Seafood CRC.
- Cotton Catchment Communities CRC.
- CRC for an Internationally Competitive Pork Industry.
- CRC for Beef Genetic Technologies.
- CRC for Forestry.
- CRC for National Plant Biosecurity.
- CRC for Sheep Industry Innovation.
- Dairy Futures CRC.
- Future Farm Industries CRC.
- Invasive Animals CRC.
- Poultry CRC.

Since the commencement of the CRC programme in 1991, there have been a number of CRCs established and de-established in various industry areas. More detailed information about all the 44 CRCs can be found in the annual publication of the *CRC Directory*. The directory provides information on each CRC's grant period and funding amounts, research focus and expertise, CRC essential partners, international collaborations, postgraduate target numbers and Chair and Chief Executive Officers' names and full contact details. It also provides information about all previous CRCs with regard to their establishment, de-establishment and the year of expiry. The latest version of this publication is *CRC Directory 2011–12* (CRC, 2012).

Most CRCs are of 7 years' duration. In 2008, the maximum term was extended to 10 years. It is generally difficult to extend a CRC. When an extension or extensions is/are granted, the aggregate duration for a CRC does not exceed 15 years. The aggregate duration may exceed 15 years only under 'exceptional circumstances'.

Australia also devotes its resources to support collaboration in international agricultural research. ACIAR is the hub through which most international collaboration on agricultural research between Australian and overseas researchers is carried out. Through ACIAR, the Australian Government channels funds to sponsor agriculture-related R&D and extension, mainly for developing countries, and hence such investment is largely foreign aid in nature. While such investment chiefly benefits recipient countries, it also benefits Australia because Australian researchers learn from the technologies they develop, from the agricultural practices of other countries and from the expertise of researchers in such countries (Pearce *et al.*, 2006).

Efforts have been made to promote further collaboration in rural research. It is argued that Australia's primary industries cannot afford a fragmented or duplicative rural research system. Research investment in primary industries needs to be focused and used efficiently, effectively and collaboratively (DAFF, 2012). Since 2005, PIMC, the Australian, state and NT governments, rural RDCs, CSIRO and universities have been working together to develop jointly a National Primary Industries Research, Development and Extension Framework.

In April 2005, PIMC endorsed the concept of 'National R with Regional D&E'. The concept recognizes that basic and strategic research (R) can be provided from a distance, with regional adaptive development (D) and local extension (E) required, improving the uptake of innovation by industry. In April 2006, PIMC agreed to a set of principles to facilitate further cooperation between agencies and industry for improving the efficiency and effectiveness of the national RD&E capability. In April 2008, PIMC agreed to the development of such a new framework. On 6 November 2009, PIMC endorsed the framework, including the statement of intent.

The new framework spans 14 primary industry sectors (including new and emerging industries) and seven cross-industry sectors. These are:

- Primary industry sectors: beef, cotton, dairy, fishing and aquaculture, forests, grains, horticulture, pork, poultry, sheepmeat, sugarcane, wine, wool and new and emerging industries.
- Cross-industry sectors: animal biosecurity, animal welfare, biofuels and bioenergy, climate change and variability, food and nutrition, plant biosecurity and water use in agriculture.

To date, PIMC has endorsed RD&E strategies for animal welfare, beef, biofuels and bioenergy, climate change and variability, cotton, dairy, fish and aquaculture, forest and wood products, grains, horticulture, new and emerging industries, pork, poultry, sheepmeat, sugarcane, water use in agriculture, wine and wool. Other industries are working progressively on their strategies for endorsement by PIMC in the near future (DAFF, 2012).

Through implementing this framework, it is expected that there will be a more coordinated and collaborative approach to rural RD&E and that national research capability will be focused and used efficiently and effectively to achieve the best outcome and uptake by primary industries. This is a very innovative and forward-looking approach. However, how effective such a framework will be is yet to be seen.

In the research community, comments about the strong emphasis on collaboration for conducting research projects through CRC arrangements and the new framework seem to be mixed. It is generally agreed that the idea to promote collaboration is beneficial. However, there have been concerns that too strong a focus on collaboration may increase the costs of collaboration and reduce competition. When commenting on the CRC model, Kerin (2010) said that, 'the Cooperative Research Centres cannot represent long-term research effort for agriculture'. With regard to the likely further boosted collaboration under the new framework,

Kefford (2011) points out that, 'collaboration has its own risks and it can go too far'. According to Kefford, collaboration is costly to establish and sustain. It also tends to lead to increased system overheads and increased need for governance, hence reduced efficiency of resource use. Kefford (2011) also argues that too much consolidation reduces competition and reduces diversity of ideas and innovation. Too strong a focus on either collaboration or competition is simplistic and hence there is a need for balance between collaboration and competition.

7.4 Agricultural Extension

Agricultural extension is crucial to the ultimate adoption of R&D outcomes. Without adequate investment in, and arrangement for, extension, the contribution of R&D to improving agricultural productivity will be compromised. This section highlights how agricultural extension is carried out in Australia.

Extension can take various forms, from the dissemination of general information on new technologies to more specific 'how to' sessions for groups of primary producers, through to one-on-one services tailored to an individual producer's particular circumstances.

Historically, extension services in rural industries were provided mainly by state and territory governments' departments of agriculture, often on a producer-specific basis, with some work also undertaken by CSIRO and universities. In recent years, the funding and delivery of extension has changed considerably. Provision of extension services by state and territory governments has changed from one-on-one activities to group extension activities, and funding has been reduced. As noted earlier, extension services provided by universities have also reduced sharply. As a result, the composition of extension service providers has changed over the past decades and many new providers have come into play. Most notably, the private sector has played an increasingly major role in the provision of extension services.

All of the RDCs are also involved in extension activities that help disseminate their research outcomes by working with extension groups, facilitating the conduct of farmer workshops, funding for demonstration farms and disseminating their research publications. Grower groups have also become involved increasingly in disseminating research results. The Kondinin Group and Birchip Cropping Group are two notable examples. There is also joint public and private investment in extension programmes. For example, the Victorian DPI, in partnership with Dairy Australia, has established the Dairy Extension Centre.

There has been an increased emphasis on extension in Australian Government programmes in areas such as conservation and sustainability. For example, the Fitzroy Basin Association (via the *Caring for our Country* programme) provides training and technical support to landholders on monitoring, managing and improving land and water quality. The National Adaptation and Mitigation Initiative, a joint investment between DAFF's Climate Change Research Program and the GRDC, aims to demonstrate on-farm climate variability adaptation measures.

Subsequently, extension arrangements in Australia today are diverse and complex. According to DAFF, extension today is a maze of different providers and access points, through private consultants, agribusiness and input suppliers, local grower groups and public information obtained through the Internet, conferences, demonstrations, workshops and publications. The result is a set of complex communication and delivery channels through which information, knowledge, new learning and ideas flow both ways (DAFF, 2010).

Several extension practices worth particular mention include the extension services provided by diverse providers at various kinds of shows, by ABC's rural radio programmes and ABC's Landline TV programme. In the author's view, these are highly effective extension approaches.

7.4.1 Shows

In Australia, almost all cities and larger rural towns have their annual shows. In addition to this kind of general show, there are also other kinds of shows organized specifically for the rural communities, such as national field days and local machinery shows. These shows give extension service providers the opportunity to promote their services through various exhibitions. Many extension providers make use of this opportunity, including both the public providers, such as the departments of the Australian Government, state and territory governments, and private providers, such as agricultural consultants, farm input producers and various other agribusiness service providers. In this way, extension services are exposed to potential users in a very relaxed atmosphere. The general public also benefit from increased exposure to rural issues.

7.4.2 ABC radio and TV programmes

The Australian Government funds rural programmes through the Australian Broadcasting Corporation, on the radio and on TV. The ABC radio provides two major rural programmes designed specifically to serve rural industries, one early in the morning a little after 6.00 am (lasting for about 20 min, chiefly on local rural issues) and the other at lunchtime for about 50 min between 12.00 noon and 1.00 pm during work days (on both local and national rural issues). In these programmes, up-to-date market information, changes in rural industries, experiences and lessons from other rural producers and any other important issues that affect rural operations are reported. Landline is a once-a-week TV programme that typically is broadcast on Sundays at lunchtime. In this programme, more important issues that have broader impacts or longer-term implications on rural industries are discussed in greater depth. All these radio and TV programmes have been most popular with rural producers. The general public also find such services provided by ABC most valuable to them to keep them informed about what is happening in rural industries, both at home and abroad.

Other methods of extension include rural magazines and newspapers and, today, the Internet as well. Private media companies run weekly rural newspapers in each state, such as *The Land* in NSW, *Queensland Country Life* in QLD and *Stock & Land* in VIC. They also publish rural magazines. Increasingly, RDCs also provide such services through printed media, i.e. magazines and newspapers. Examples are *Ground Cover* by GRDC, *Rural Diversity* by RIRDC and *Feedback* by MLA. State and territory departments of agriculture provide information services through printed media as well, though this is declining. All of them provide information services to the rural community through the Internet; most of them can be accessed free of charge. Access to such information provided by some private consultants on the Internet may require a subscription fee (for example, the Analysing Agriculture newsletter produced by John Chudleigh of NSW at AnalysingAgriculture.com). In short, today Australian farmers have access to information, knowledge and new ideas through a set of complex communication and delivery channels.

7.5 Agricultural Education

Agricultural education crucially affects the performance of agricultural RD&E. It provides qualified personnel to engage in agricultural RD&E. Agricultural graduates, when working on farms, have a better chance to understand and adopt new R&D outcomes. This section highlights briefly the provision of agricultural education in Australia.

In Australia, agricultural education is provided in both secondary and tertiary education. Education in agriculture at the secondary level is funded by the Department of Education in each state. Secondary schools do not provide comprehensive vocational training in agriculture; rather, the emphasis is on general education through offering some agriculture-related subjects. Some secondary schools in rural areas provide minor specialization in agriculture, as do a small number of secondary schools in metropolitan areas. Also, some secondary schools, in both rural and urban areas, have the word 'agricultural' or 'rural' in their names: for example, the Canobolas Rural Technology High School in Orange, NSW, and James Ruse Agricultural High School in Sydney.

Agricultural education at the tertiary level is funded mainly by state or federal governments. Three types of institutions provide tertiary agricultural education: technical and further education (TAFE) colleges, agricultural colleges and universities. Together, they provide agricultural education at varying levels to suit the needs of the farming community.

7.5.1 TAFE

Each state has a TAFE authority providing a variety of forms of technical education and training in various industries, including agriculture. Courses related to agriculture offered by TAFE have a strong vocational component including wool classing, farm mechanics, farm welding, landscaping, etc. TAFE also offers courses in general agriculture, horticulture, fruit growing, pig and poultry farming, farm record keeping, and so on. TAFE courses can be undertaken on either a part-time or full-time basis. To date, the highest degree TAFE can offer is a diploma.

7.5.2 Agricultural colleges

Agricultural colleges have played a significant role in providing education in agriculture for over a century, with the first, Roseworthy Agricultural College, being established in 1883. There were once almost 20 agricultural colleges in Australia: all of them but one were publicly funded. Marcus Oldham in VIC is the only private agricultural college. These colleges played a major role in the education of farmers, extension workers and agricultural administrators. During the 1980s and 1990s, important changes occurred to many of the publicly funded agricultural colleges: they were either closed or became part of a university under John Dawkins' (the then Minister for Education) education reforms. Closure of such colleges is still continuing. The most recently closed college was the Muresk Institute of Agriculture in WA, which was founded in 1925. Today in Australia, there are only a very small number of publicly funded agricultural colleges. On the other hand, as a privately funded agricultural college, Marcus Oldham is experiencing expansion and offers a number of courses including farm management, agribusiness and horse business management, at various levels, such as advanced diploma, diploma, bachelor, postgraduate certificate and postgraduate diploma.

7.5.3 Universities

Degree courses in agriculture are available in at least one university in each state. Such universities also provide postgraduate facilities leading to postgraduate diploma, Master degrees and Doctor of Philosophy or Doctor of Science degrees. Several universities famous in providing agricultural education and training include the University of Sydney, the University of New England, the University of Queensland, the University of Western Australia, Murdoch University, the University of Adelaide and the University of Melbourne.

While the above institutions provide agricultural education, some other organizations, such as RDCs and private companies, also invest in education through providing scholarships for students to undertake agriculture-related courses. Some of them fund facilities that provide agriculture-related education to the general public. For example, Dairy Australia jointly funds the National Centre for Dairy Education Australia, and the Cotton RDC invests in education indirectly via its funding for the Cotton Catchment Communities CRC (PC, 2011, p. 34).

7.6 Private Investment in Agricultural RD&E

According to Keogh (2011), in most developed nations, the private sector is playing an increasingly important role in agricultural R&D. The share of total agricultural R&D investment sourced from the private sector averaged 54% for OECD nations in 2000. Private investment in agricultural R&D in Australia, however, has been small and has increased only gradually in recent years (Campbell, 1980,

p. 143; Jarrett, 1990; Mullen, 2007). According to Mullen (2007), private funding in agricultural R&D was minimal some 20 years back. The private sector was responsible for less than 10% of total agricultural R&D, although its share in 2007 was increased to 20%.

A number of factors may be responsible for the low private investment in agricultural R&D in Australia. (i) Agricultural R&D is often very expensive. For example, the cost of developing a new crop variety can be very costly. (ii) Australia has a relatively small market in terms of demand for agricultural R&D products, offering limited commercial incentives. (iii) There are spillovers flowing into Australia from international agricultural R&D investment. (iv) Some government regulations may have constrained private investment in certain research areas such as the development of genetically modified crop varieties. (v) The non-excludability and non-rivalry characteristics of information generated by research simply mean that no private investment will be made for certain R&D activities (Campbell, 1980, pp. 143–144; Jarrett, 1990; Keogh and Potard, 2011; Mullen, 2011).

As a result of reduced public funding in some agricultural R&D in recent decades, and also relaxed government regulations on some R&D activities, private investment in agricultural R&D has been increasing slowly. According to PC's rural R&D report (2011, p. 19), there are three main sources of private funding for rural R&D in Australia.

- Industry payments to the RDCs, industry-owned research institutions (such as BSES Limited, an entity that performs some AUS$20m/year of sugar research) and state-based research organizations (such as the South Australian Grains Industry Trust and the (WA) Agriculture Produce Commission).
- Large commercial farming companies such as Auscott Limited, Clyde Agriculture, Huon Aquaculture, PrimeAg and Twynam.
- Chemical and fertilizer research companies such as BASF, Bayer, Dow, Monsanto, Nufarm, Pfizer and Syngenta (which also make large investments in rural R&D internationally).

The PC estimates that, collectively, private entities fund around 25% of overall rural R&D. This estimate is, however, believed to be too low and has been seriously challenged. Keogh and Potard (2011), based on a survey of private investment in agricultural R&D, believe the private sector has provided a considerably higher share of total funding than the Commission's estimates. The PC argues that some missing data and questionable methodologies in Keogh and Potard (2011) compromise the reliability of AFI's estimates. On the other hand, Keogh (2011) claims that the PC seems to have attempted deliberately to inflate the apparent share of public funding of rural R&D in Australia in order to support better a recommendation to reduce public funding for this activity in the future.

In spite of the arguments about the actual share of contribution from private investment into agricultural RD&E, it has been felt increasingly in Australia that the private sector's role in this area is growing and private investment is expected to rise further.

7.7 Remaining and Emerging Issues

Clearly, Australia has done well in instituting a well-devised agricultural RD&E system in the past three decades. However, some issues concerning this front remain or are emerging.

- **Declining public funding in agricultural RD&E.** Research has shown consistently that the rate of return to research in Australian agriculture is high, being in the range of 15–40% (Mullen and Cox, 1995; Mullen, 2007; Sheng *et al.*, 2011). However, it has been noted that public investment in agricultural RD&E has been declining in recent years (Mullen, 2010a, 2011).

  The decline in state-level funding for agricultural R&D has recently attracted much warranted attention. According to the PC (2011, p. 19), many submissions expressed concerns about what they perceived to be a progressive withdrawal of state and territory governments from the agricultural R&D area (see, for example, Across Agriculture, 2010; Mullen, 2010b; South Australian Farmers, 2010). The PC acknowledges that 'it has not seen any evidence that refutes this common perception'.

  The decline in public funding for agricultural RD&E is most worrisome and may undermine Australia's agricultural productivity in the longer term (Sheng *et al.*, 2010; Gray *et al.*, 2012). The literature has demonstrated that it takes a long time for the benefits of agricultural R&D investment to emerge; some believe it is likely to be 30 years or more (Mullen, 2011). Many have argued that today's satisfactory agricultural productivity growth in Australia is attributable to the relatively large investment in agricultural R&D in the 1970s and 1980s. Without such investment, today's productivity could be much lower (Mullen, 2007; Sheng *et al.*, 2011). For the same reason, the damage caused by the lack of agricultural R&D investment today will not emerge until a very long time later. Hence, being complacent about today's satisfactory agricultural productivity growth is not warranted and reducing public funding for agricultural R&D is short-sighted. Public investment in agricultural R&D in Australia remains a sensible policy option (Mullen, 2011; Gray *et al.*, 2012).
- **Recommendation by the PC to reduce public funding for RDCs.** Related to the above issue, the 2010 inquiry into the RDCs has led the PC to recommend reduced funding for RDCs – 'Over ten years, halve the current cap on the Government's dollar-for-dollar matching of industry contributions', in order to 'shift to a more appropriate balance between private and public funding responsibilities for industry-focused R&D' (PC, 2011, p. xxxvi). This has resulted in a barrage of outcry and many have argued against this recommendation. This is evidenced by the large number of submissions (being 132) made by various stakeholders in response to the draft report release, reflecting the strong interest in the community to support an adequate level of public funding in agricultural R&D (all these submissions are available online at: www.pc.gov.au/projects/inquiry/rural-research).

The current Australian Government has rejected this recommendation. However, concerns remain: the current government is a minority government and relies critically on two independents that have strong agricultural constituents. What happens when such a situation changes in the future? The Australian rural industries need to remain vigilant to argue strongly for higher-level public funding for rural RDCs. Agricultural R&D will, indeed, benefit not only those agricultural industries but also the broad community through flow-on benefits of improved agricultural practices. Further, according to Keogh and Potard (2011), the level of public R&D investment is one of the most important factors that encourages private investment in agricultural R&D.

- **Lack of attention to agricultural extension and education.** While it is fair to say that Australia has been doing well in agricultural extension and education, generally there is the feeling that, in recent years, there is reduced attention to these two areas. Many others also acknowledge this issue, including the PC (PC, 2011, p. 288).

  Extension service provided by state and territory departments of agriculture has contracted steadily in the past three decades. The farming community is generally not happy about it, although there have been no major outcries. Within the research community, some argue that as long as there is no market failure, a reduced public extension service and increased private extension services is a sensible shift (B. Malcolm, the University of Melbourne, 7 October 2010, personal communication). Others, however, argue that the scope of the service may be compromised as the private sector will only provide services that can help them earn a profit (R. Murray-Prior, Curtin University, 7 February 2012, personal communication).

  Driven by commercial interests, the private sector will only engage in profitable extension activities. There are other extension services that, though commercially not profitable, can be vitally important for the rural community and can also generate significant positive externalities.

  In the case of agricultural education, it seems there is a lack of long-term policy. Rather, to a great extent, the provision of agricultural education facilities has been left to the market. When the demand is low, such facilities have been allowed, or forced, to close. Over the past two decades, most publicly funded agricultural colleges disappeared. Many agriculture-related courses provided by universities also closed or shrank. This has led to the loss of agricultural education facilities and expertise. It can be difficult to rebuild such facilities and recruit staff when needed in the future.
- **The cut of training for vocational purposes.** Those previously publicly funded agricultural colleges specialized in the vocational training of farmers and young people who planned to work on the land. Under Dawkins's plan, many of them became part of a university. This forced marriage seriously undermined their ability to provide vocational training, for at least two reasons: (i) universities are less interested in providing vocational training; and (ii) as part of a university, academics are required to perform according to university performance criteria, such as producing refereed publications. Australia's capability to provide vocational agricultural training has been compromised.

7.8 Concluding Comments

Investment in agricultural RD&E has contributed importantly to Australia's attainment of a reasonable level of agricultural productivity growth. Across Australia, RD&E is funded and carried out by a complex and diverse web of research providers and investors with strong interconnections. Public investment accounts for a much larger share in the total investment in agricultural RD&E, but is declining. Private investment, which accounts for a relatively smaller share, is on the increase, however.

Out of the public investment, a relatively larger share of the funds from the Australian Government is provided to the RDCs and CSIRO, followed by universities and other departmental programmes. CRCs are also allocated funds from the Australian Government, but at a much smaller share.

The 15 rural RDCs are a unique partnership between the Australian Government and the rural industries. Their focus is on expanding Australia's rural R&D effort, improving industry effectiveness and efficiency by investing in high priority areas and encouraging uptake of research results to improve international competitiveness and sustainability. They commission and manage targeted research, and foster uptake and adoption based on the identified needs and priorities of both industry and the Australian Government. The Australian Government provides dollar-for-dollar matching of industry expenditure on R&D up to a limit of 0.5% of each industry's GVP.

The RDC model today is a mix of statutory and industry-owned companies. Together, they cover virtually all of the agricultural industries. The co-investment model of industry and government funding has been an efficient and effective means of managing the Australian Government's investment in rural R&D. The CRRDC serves as a forum for the 15 RDCs which assists in the coordination of their activities.

Collaboration in agricultural RD&E has been promoted enthusiastically in Australia. In addition to various collaborative activities coordinated by the RDCs, several CRCs are engaged mainly in collaborative research. ACIAR is engaged in international agricultural research collaboration. Currently, a much wider and more coordinated collaborative research framework, the National Primary Industries RD&E Framework, is being promoted. Collaboration is essential to use limited funds most efficiently. However, excessive collaboration can be costly, reduce necessary competition and reduce diversity of ideas and innovation. Whether the new framework will be efficient and thus successful is yet to be seen.

In Australia, public provision of agricultural extension has been declining. Extension services provided by the private sector are on the increase. Australia has been doing very well in providing extension services to the rural community through radio, TV, various new telecommunication and information technologies, and community shows. The provision of vocational agricultural education by public-funded providers has declined. The demand for enrolments in university agricultural courses has also declined. Australia is likely to face a shortage of workers with vocational skills and university-level training in agriculture.

Agricultural research, development, extension and education have been central to increasing industry productivity and ensuring sustainability in Australia. Continued investment at an adequate level will be crucial for future agricultural success. The recent declining trends in public investment in agricultural RD&E have been a concern and must be arrested.

For both developed and developing countries, Australia's practice in investing in agricultural R&D is of some relevance. Clearly, the importance of public investment in agricultural R&D should not be overlooked. Public spending is essential (although the share of private investments in agricultural R&D in some developed countries is relatively high). Private investments are made for pursuing profits. Some agricultural R&D activities may not generate immediate and measurable benefits and thus do not attract the private sector, but they are of long-term significance. Public investments have to be made in such areas. In developing countries, due to lack of private investment, public expenditure in agricultural R&D is even more important.

It must be noted that no amount of private investment can overcome the fundamental market failure associated with the provision of R&D. R&D has characteristics of non-rivalry and non-excludability. These characteristics make some degree of market failure inevitable, emphasizing the importance of public investment in agricultural R&D.

Australia's RDC model is of relevance to any other country where co-investment has not been trialled. However, one pre-condition must exist in a country if such a model is to be successful. That is, good governance of public money must be present to ensure the levy contributions collected from farmers are actually used for farmers' benefits. The management and use of the collected funds must be transparent. Those who are involved in managing and spending the funds, i.e. the staff of funding bodies and researchers, must be held accountable for the use of the funds. Mechanisms must exist to prevent the levies collected from farmers from becoming prey of corrupted officials, or being misused or abused by funds-managing staff and researchers. Many developing countries do not have strong institutions in governing the use of public money. Caution would have to be exercised should any of them contemplate the adoption of such a co-investment model.

In many countries, public funding for agricultural R&D is not abundant and is, in many cases, declining. Making use of limited funds most effectively and efficiently is therefore important to offset to some extent the impacts of reduced funding. Collaboration can be a useful approach. It helps to avoid repetition and duplication. In many developing countries, collaboration among researchers is weak. Where appropriate, some requirements may be attached to the granting or allocation of public funds to induce collaboration. However, excessive collaboration should be avoided.

The use of TV, radio and printed media to disseminate information to farmers for extension and education purposes is gaining popularity in some developing countries. These methods have been used extensively in Australia. They are most cost-effective. Increased efforts to use such public media, and where possible some other new information and communication technologies and public gatherings, would be duly rewarded for many developing countries.

For developed countries, the share of private investment in agricultural R&D is relatively high; perhaps because the agricultural sector is large and there is strong protection of intellectual property. However, even in these economies, there will still be significant market failure in the provision of agricultural research, especially where there are few mechanisms to allow farmers to fund R&D collectively. These problems are even larger in small developed economies and in developing economies, and so the share of agricultural R&D funded by the public sector will necessarily be larger.

References

Across Agriculture (2010) Across Agricultural Submission: Productivity Commission Inquiry into Rural Research and Development Corporations, 28 June 2010, submission number 116 (http://www.pc.gov.au/__data/assets/pdf_file/0004/99445/sub116.pdf, accessed 15 August 2011).

BSES (2010) Submission to the Productivity Commission Inquiry into the Australian Government Research and Development Corporations Model. BSES Limited, 24 June 2010, submission number 42 (http://www.pc.gov.au/__data/assets/pdf_file/0005/99347/sub042.pdf, accessed 15 August 2011).

Campbell, K.O. (1980) *Australian Agriculture: Reconciling Change and Tradition*. Longman Cheshire, Melbourne, Australia.

CRC (2012) *CRC Directory 2011–12*. CRC programme of the Department of Innovation, Industry, Science and Research, Canberra.

CRRDC (2010) Submission to the Productivity Commission Inquiry into the Australian Government Research and Development Corporations Model. Council of Rural Research and Development Corporations, 3 July 2010, submission number 128 (http://www.pc.gov.au/__data/assets/pdf_file/0009/99531/sub128.pdf, accessed 15 August 2011).

CRRDC (2012) Welcome to the website of the Council of Rural Research & Development Corporations (http://www.ruralrdc.com.au/Page/Home.aspx, accessed 7 April 2012).

CSIRO [Commonwealth Scientific and Industrial Research Organization] (2011) About CSIRO (http://www.csiro.au/org/About-CSIRO.html, accessed 22 July 2011).

DAFF (2010) Submission to Productivity Commission Inquiry into the Australian Government Rural Research and Development Corporations Model. Department of Agriculture, Fisheries and Forestry, 4 August 2010, submission number 156 (http://www.pc.gov.au/__data/assets/pdf_file/0018/100683/sub156.pdf, accessed 15 August 2011).

DAFF (2012) National Primary Industries Research Development and Extension Framework (http://www.daff.gov.au/agriculture-food/innovation/national-primary-industries, accessed 31 March 2012).

Gray, E.M., Sheng, Y., Oss-Emer, M. and Davidson, A. (2012) Agricultural productivity: trends and policies for growth. *Agricultural Commodities* 2, 166–179.

Jarrett, F.G. (1990) Rural research organisation and policies. In: Williams, D.B. (ed.) *Agriculture in the Australian Economy*, 3rd edn. Sydney University Press, Sydney, Australia, pp. 82–96.

Kefford, B. (2011) What is changing in primary industries research, development and extension in Australia and why? A keynote speech delivered at the 2011 Australian Agricultural and Resource Economics Society, Melbourne, 8–11 February 2011.

Keogh, M. (2011) Private sector investment in agricultural research and development in Australia. *AFBM Journal* 8, 14–19.

Keogh, M. and Potard, G. (2011) Private sector investment in Agricultural R and D in Australia. Research report, Australian Farm Institute, Sydney, Australia.

Kerin, J. (2010) What policy framework would I now establish for agricultural research, development and extension, if I were still Minister for Agriculture, Fisheries and

Forestry. A speech given to the University of Melbourne, 5 May 2010 (http://www.landfood.unimelb.edu.au/info/seminars/2010/johnkerin%20-%20DLS2010.pdf, accessed 16 November 2010).

Mullen, J.D. (2007) Productivity growth and the returns from public investment in R&D in Australian broadacre agriculture. *Australian Journal of Agricultural and Resource Economics* 51, 359–384.

Mullen, J.D. (2010a) Trends in investment in agricultural R&D in Australia and its potential contribution to productivity. *Australasian Agribusiness Review* 18, 18–29.

Mullen, J.D. (2010b) Submission to the Productivity Commission Inquiry into the Australian Government Research and Development Corporations Model. 9 December 2010, submission number DR172 (http://www.pc.gov.au/__data/assets/pdf_file/0017/103661/subdr172.pdf, accessed 15 August 2011).

Mullen, J.D. (2011) Public investment in agricultural research and development in Australia remains a sensible policy option. *AFBM Journal* 8, 1–12.

Mullen, J.D. and Cox, T.L. (1995) The returns from research in Australian broadacre agriculture. *Australian Journal of Agricultural Economics* 39, 105–128.

Nossal, K. and Sheng, Y. (2010) Productivity growth: trends, drivers and opportunities for broadacre and dairy industries. *Australian Commodities* 17, 216–230.

PC [Productivity Commission] (2011) Rural Research and Development Corporation. PC Inquiry Report, No 52, Canberra.

Pearce, D., Monck, M., Chadwick, K. and Corbishley, J. (2006) *Benefits to Australia from ACIAR-funded Research*. Australian Centre for International Agricultural Research, Publication Code: IAS039, Canberra.

Sheng, Y., Mullen, J.D. and Zhao, S. (2010) Has growth in productivity in Australian broadacre agriculture slowed? ABARE Conference paper 10.01, presented to the Australian Agricultural and Resource Economics Society, 10–12 February 2010, Adelaide, Australia.

Sheng, Y., Gray, E.M. and Mullen, J.D. (2011) Public investment in R&D and extension and productivity in Australian broadacre agriculture. ABARES Conference paper 11.08, presented to the Australian Agricultural and Resource Economics Society, 9–11 February 2011, Melbourne, Victoria, Australia.

South Australian Farmers Federation (2010) Submission to the Productivity Commission Inquiry into Rural Research and Development Corporations. South Australian Farmers Federation, 25 November 2010, submission number DR199 (http://www.pc.gov.au/__data/assets/pdf_file/0004/103936/subdr199.pdf, accessed 15 August 2011).

8 Innovative Farming Community

It is pointed out in Chapter 6 that farmers in Australia need to be entrepreneurial. They also have to be innovative, being able to invent new tools or methods to improve their operations. Collectively, Australian farmers have been very innovative. In return, they are rewarded with increases in farm output and profitability. Their activities are also widely recognized and rewarded by various industry and government schemes that encourage and foster farmers to be innovative. The next section highlights the need to be innovative in Australian farming. Then, presented in Section 8.2 are several examples that demonstrate how innovative Australian farmers have been. Section 8.3 shows how farmers' innovation is fostered, recognized and rewarded.

8.1 The Need to be Innovative

Definitions of innovation vary. The *Australian Macquarie English Dictionary* defines innovation as 'introducing new things or methods'. This simple definition serves the purpose of discussions in this chapter. Hence, in the context of Australian agriculture, any new tools or new management methods invented or introduced by Australian farmers are regarded as innovation.

There was no choice but to be innovative in order to farm in Australia. There were several major reasons why this was the case: (i) lack of indigenous agriculture; (ii) geographical isolation; and (iii) survival.

8.1.1 Lack of indigenous agriculture

In his chapter that describes innovations in Australian agriculture, Donald observes:

> When the first wheat was sown in Sydney in 1788, Australia's agricultural resources could scarcely have been more meager – an unknown climate, the miserably poor soil of the Sydney area, convict labour and inexperienced husbandmen. It is not surprising that the first wheat crop was a failure nor that the next gave only a fourfold return of the seed sown.
>
> (Donald, 1982, p. 55)

The Aborigines in Australia at that time were collectors and hunters. There was a lack of any indigenous agriculture. Those pioneer settlers had to start from scratch and develop new ways to farm the Australian land. They had to discover plants and animals, introduced from overseas, that would suit the Australian soil. Few of the many native plants have entered Australian agriculture, because native plants have adapted to the low fertility of most of the soil and are unable to respond to improved soil conditions. Also, few of the many native animals have entered the Australian pastoral industries due to the prior domestication and known productivity of European species and breeds.

In addition to the selection of plants and animals, pioneer settlers and their successors have also had to be innovative on many other fronts, such as soil improvements, soil conservation, the control of pests and the use of modern chemicals. A useful summary of the major innovations in earlier years that led to significant breakthroughs in Australian agriculture can be found in Donald (1967, 1982).

8.1.2 Geographical isolation

Two centuries back, the distance between Australia and any other major civilizations was enormous. If any farming equipment was needed, relying on imports from elsewhere was expensive and often required further adaptation. When Australia's particular environment conditions (such as dry with limited and erratic rainfall, and poor soil) are also taken into consideration, then not many environments elsewhere could lend Australia many effective farming practices. Hence, not only did distance make imports less practical, but also the peculiar environment required the earlier settlers to innovate new equipment and farming methods that suited Australian conditions.

Today, although many farming facilities that suit Australian conditions have become readily available, geographical isolation has meant that farmers still need to be capable of innovating. Some Australian rural properties are several hundred, or even thousand, square kilometres large and are in very remote areas. When unexpected circumstances occur, for example machine breakdown, waiting for part replacements to arrive can be very time-consuming, or sometimes the situation can even be life-threatening. One may have no choice but to find alternative ways to make things work. In rural areas, there is no shortage of stories about how farmers work out some magic ways to fix things that are often beyond imagination. One example follows.

Jock MacDonald, now retired, used to manage a cattle station in the Gulf of Carpentaria area in the tropics of Far North QLD. On one occasion, the bottom bung of the differential of his Holden FC utility was lost. The nearest place to get a replacement was some 200 km away. He was by himself. Waiting for assistance was not an option, as the road might not be used by anyone for several days. It is life-threatening if one runs out of food and water in the hot and humid tropics. Jock found a piece of gidgee wood (a type of acacia, very hard wood) and he made a wooden bung to get his vehicle moving to reach the nearest place for proper replacement.

8.1.3 Survival

In the earlier days of European settlement, given the Australian land was vastly different from what the settlers were familiar with at home, they had to find new ways to farm in order to survive physically (Clark, 1992, pp. 23–27). Today, they have to be innovative in order to survive economically. They have continuously to improve the way they farm so that they can cope with pressures from competitive markets, climate variability, environment protection, and so on.

Over the past 220 years, innovations by farmers have been supplemented and surpassed progressively, but never replaced completely, by contributions from public and private research. The innovative culture remains strong among the Australian farming community. Many farmers continue to discover or invent new ways to farm. Some recent examples are provided in the next section.

8.2 Some Examples

8.2.1 Example 1: Optimizing farm operations to increase income and reduce costs

John Mieglich runs his own small dairy farm in partnership with his brother in Charleston, a small country town that is about 25 km east of Adelaide, SA. John, his brother and an employee work together on the farm, which is about 250 acres in size, carrying about 250 animals. John's wife, Pam, also provides some casual assistance. The scale of the operation is relatively small. The price of land in the region has become very expensive and they are not in a position to expand their farm. To generate an income sufficient to support the lives of the four people requires them to optimize their work on the farm. Instead of producing milk for supplying to the mass market, they specialize in producing premium-quality milk to supply to some niche markets, hence commanding a premium price. Otherwise, there would be no way to support the four financially. Despite this market segmentation, they still have to be very vigilant to keep their costs under control.

They use a variety of innovative approaches to keep their costs under check. Some of these include:

1. Where possible, programme electricity use during off-peak times to benefit from lower tariffs. For example, irrigate the pasture overnight (which also

reduces water evaporation, thus reducing water costs as well); make ice overnight for use during the daytime to chill milk.

2. Install a soil moisture probe to monitor the soil moisture to aid irrigation decisions, thus saving water, and electricity as well.
3. Use labour efficiently by arranging activities in ways that reduce or avoid unproductive activities.
4. Make use of the wastes generated by the animals to improve the pastureland, thus reducing the cost of buying fertilizers and also avoiding the costs of removal.
5. Use different crops for the pasture according to soil conditions and season, to save water and increase feed.
6. Use different forms of fertilizer according to the seasonal conditions, to reduce leaching. In cold temperatures, it takes longer for urea to be taken up by the plant. If solid fertilizer is applied to a paddock when it is cold and wet, it may leach to the subsoil, where the plant cannot use it. In such situations, liquid fertilizer is applied to the crops so that the leaves of the crop will absorb the fertilizer immediately.

8.2.2 Example 2: Passion for cross-breeding

Jim Sullivan is a pioneering farmer who, in the 1960s, took his young family from his NSW home at Coonamble where he grew wheat, to the NT to fatten cattle, and who later trialled rice, bananas, melons and other crops with success. His initial base was Wandinya, a pilot farm. Later, he and his young family moved around the NT and then settled at Cave Creek Station, 10 km from Mataranka, a small town that is 100 km south of Katherine.

Over 70 years old, Jim, officially, has retired. Yet, his passion for breeding cattle remains. He is still working on trials: to cross-breed Brahman with a small herd of high-fertility pure African Boran cattle that he bought in from Toowoomba in QLD in 2003. He calls the trials 'Jim's folly' – something else for him to do in between his other jobs of doing the lick drops to the herd twice a week, helping maintain the fencing and other duties as they arise.

The cattle industry in the NT has taken huge steps forward since Jim arrived in 1964. The Brahman breed has been introduced and cattlemen have discovered the value of supplementary feeding to make up for lack of nutrients in native pastures. Live export shipments have started and TB has been eradicated. Jim believes strongly that one main thing left for the industry to do is to lift the fertility levels in the herds.

Brahman and Boran are both Zebu-type cattle that originated in India. The Boran was introduced into Africa about 2000 years ago. The Boran people selected these two breeds for milk production because they had high fertility capabilities. However, people in the industry do not have a high regard for Boran. Not surprisingly, not all people support Jim's venture.

Nothing can stop Jim's passion for his trials. He argues that when CSIRO brought in Brahman for trials in the 1930s, people did not have a high regard for them either. It took some 20 years for Brahman to be accepted. In the earlier days, people thought that no breed could be better than the British breeds. But now, people know that CSIRO was right. Jim is confident that his trials will be successful. He foresees that in the not-too-distant future, there will be a lot of Brahman and Boran cross-bred cattle on his own property, Cave Creek, and eventually throughout the NT.

8.2.3 Example 3: Managing pasture production to reduce feed costs

Wes Brown manages a pastoral property owned by his parents near Cargo in the Central West Tablelands of NSW. Rainfall on the property is around 800 mm per annum. However, the distribution of rainfall is uneven. Droughts also hit. Animal feeds, either buying from others or storing on the farm, add to cost. Wes has developed his philosophy of pasture management to minimize the need to buy or store feeds. The focal point of his pasture management approach is to choose plants that maximize photosynthesis. To achieve this, he chooses: (i) where possible, plants with the largest leaves; and (ii) plants that can be sown and will mature at different points of time, with varying drought resistance. In this way, he increases the output of his pasture crops and his animals can graze directly from the paddock for an extended period, reducing feed storage. Wes's smart pasture management approach has benefited from his education from Orange Agricultural College and a course on plant physiology he is currently doing part-time through Charles Sturt University.

8.2.4 Example 4: Sediment trapping

Topsoil contains fertilizers and nutrients needed to boost crop growth. It is therefore essential to reduce soil erosion and nutrient losses from farms. Not only does this improve farm productivity and profitability but it also reduces pollution of the environment. A group of farmers from the Silkwood Drainage Board near Innisfail in Far North QLD, Australia's major sugarcane-producing region, investigated ways to reduce soil erosion and nutrient losses from cane farms.

To reduce soil loss, the group installed sediment traps at the end of paddocks, in irrigation channels and in creek beds, to capture soil and prevent sediment entering waterways or local catchments. The results were impressive, and the traps reduced soil sediment movement significantly and created positive economic and environmental benefits. For their innovative achievements, the group won the 2011 Sugarcane Grower Group of the Year Award, which was one of nine categories linked to the Australian Farmer of the Year Awards, hosted by the Kondinin Group and ABC. The group has published a booklet, *Sediment Trapping Field Guide*, to assist other farms across the cane industry in designing and installing sediment traps to reduce the impact of soil sediment movement (Grower Group Services, 2012; SRDC, 2012a).

8.3 Encouraging, Fostering and Rewarding Farmer Innovation

Although many passionate farmers in Australia are intrinsically creative in their ways of doing things on the farm, the presence of institutions that facilitate farmers' pioneering efforts is equally important in building Australia's innovative farming community. Such institutions include rural RDCs, government departments, professional bodies and farmer organizations. These organizations encourage farmers to innovate, provide resources to foster farmers' efforts to innovate and recognize farmers' achievements in innovation by rewarding them in various ways.

For example, to encourage and foster farmers to be inventive, the Sugar Research and Development Corporation (SRDC) developed a funding programme specifically to encourage and foster farmers to work together to develop on-farm innovative activities. The programme is called Grower Group Innovation Projects (GGIPs) (SRDC, 2012b). GGIPs help a group of growers to build their capability for innovation, by conducting their own research into more profitable and environmentally sustainable sugarcane farming systems. GGIPs first began in 2005. Since then, over 50 projects have been funded. These projects are located across Australia's major cane-producing areas in NSW and QLD.

A group of growers of any size, from any Australian sugarcane region, can submit an application to attract part or full funding support from SRDC for their project. Proposals generally need to demonstrate that a group has access to the appropriate skills to design and conduct a research activity, and to interpret and communicate the results. Depending on the scope of individual projects, SRDC provides funding of up to AUS$80,000 per project over a term of up to 3 years.

Successful grower groups then undertake on-farm trials or machinery modifications. Such groups report back to SRDC on the results of the project through regular milestone reports. Groups also need to undertake communication activities such as field days, bus trips, media articles, etc., to promote their results throughout the sugar industry.

Previous topics funded under GGIPs include row spacing comparisons, investigating alternative legume crops, utilizing compost as an alternative fertilizer source, developing a precision mill mud applicator and many more. The sediment trapping investigation as given in the previous section was another example that was supported by the GGIPs programme.

Various other organizations also provide encouragement to farmers to foster their innovation activities similar to the support from SRDC. For example, GRDC supports a large number of grower groups nationally (GRDC, 2012). One group supported by GDRC is Central West Farming Systems (CWFS), which is based at Condobolin in central west NSW. CWFS has been a platform for innovative extension and experimentation in the farming sector of the region. This has enabled local farmers to undertake experimentation in their own environment. CWFS has a slogan: Farmers Advancing Research (CWFS, 2012).

Not only do institutions encourage and foster farmers' innovation activities, they also provide public recognition of their innovative achievements. In the case of SRDC, it provides Grower Innovation Awards to profile and acknowledge those farmers with a real commitment to initiate change in the sugarcane industry. The awards recognize the contribution made by either a group of growers or an individual grower. Winners are also provided with financial support to help them continue their on-farm research or to invest in a professional development activity (SRDC, 2012c).

The annual SRDC Grower Innovation Awards offer four award categories:

- **Grower Group Innovation Award:** recognizes the outstanding achievements of a grower group that has completed an innovative research project.
- **Grower Group Progress Award:** recognizes the successful progress a group of growers has achieved during the preliminary stages of a project not yet complete.
- **Grower Service to Industry Award:** recognizes an individual cane grower who provides outstanding leadership and service to the sugarcane industry.
- **Project Supporter Award:** acknowledges the development and extension services that an individual has provided a group of growers to complete a project.

Similarly, various other organizations have their own award programmes to recognize the innovation achievements of farmers. At the 2010 National Congress of the NFF, two Innovation in Agriculture Awards were conferred (NFF, 2011, pp. 26–27). John, Bryan and Terry Granshaw – third-generation cane growers from Dalbeg in QLD – took the Sustainability Award, which recognized the coexistence of boosting farm production while ensuring environmentally sustainable outcomes. The Granshaws were among the first cane growers to stop burning and embrace green cane harvesting 20 years ago. Since then, through adopting various techniques innovatively, they have reduced fertilizer and chemical use by 60%, reduced soil compaction and reduced water use from 10-day to 28-day cycles, all while improving soil moisture and organic matter levels to lift production levels (NFF, 2011, p. 26).

Phil and Lynda Snowden from Tocumwal in NSW won the New Technology Award, which recognized those farmers that had instigated, adapted or seized upon technological advances and employed them with great effect on the farm. The Snowdens were rewarded for a revolutionary invention, Hay Caps, for storing hay safely and efficiently. Haystacks are typically high and dangerous to cover using traditional tarpaulins. The Snowden's invention has made covering the large, square bales of hay safe, quick and economical. Fitted safely atop hay bales at ground level, covers are placed on the bales and secured with pegs. The tractor picks up and stacks the bales, ensuring the covered bale is on top. As the haystack is built, the folds of the Hay Caps butt up against each other, creating a roof, ensuring full coverage with no gaps between the bales, even if the stacking is a little rough. The cover is made of food grade, recycled plastic. It is flexible, vermin-proof and durable for up to 10 years of use, unlike tarpaulins, which have to be replaced after one or two seasons (NFF, 2011, p. 27).

8.4 Concluding Comments

To survive and prosper, farmers everywhere are innovative. Like many farmers elsewhere, Australian farmers have been very innovative in their farming. Facilitating institutions have played an important role in fostering a strong innovative culture in the Australian farming community. Indeed, being innovative has become a valuable inheritance that many young farmers care to carry on. Today, there are policies from government departments and industry bodies that preserve this valuable inheritance and promote and encourage all kinds of innovative efforts by farmers. Farmers' innovations, together with innovations from other R&D providers, will continue to play a major role in advancing Australian agriculture further.

For any ordinary person in society to be innovative requires humans to have the freedom to think and express themselves. Lack of this freedom discourages and even destroys one's ability to innovate. To encourage farmers, just like any other people, to be innovative, freedom of thought and expression is essential. When this is lacking, as in dictatorship societies, it is hard for anyone, including farmers, to be innovative.

For farmers in poor developing countries, lack of resources may limit innovation to some degree. Innovation may require some support. The farmer innovation group model used in Australia can be referred to when providing resources to those hopeful farmer innovators.

References

Clark, M. (1992) *A Short History of Australia*. Penguin Books, Ringwood, Victoria, Australia.

CWFS [Central West Farming Systems] (2012) Farmers advancing research (http://www.cwfs.org.au/, accessed 4 April 2012).

Donald, C.M. (1967) Innovation in Australian agriculture. In: Williams, D.B. (ed.) *Agriculture in the Australian Economy*. Sydney University Press, Sydney, Australia, pp. 57–86.

Donald, C.M. (1982) Innovation in Australian agriculture. In: Williams, D.B. (ed.) *Agriculture in the Australian Economy*. Sydney University Press, Sydney, Australia, pp. 55–82.

GRDC [Grains Research and Development Corporation] (2012) Grower groups (http://www.grdc.com.au/director/events/grdcpublications/growergroups.cfm, accessed 4 April 2012).

Grower Group Services (2012) Sediment trapping field guide (http://www.growergroupservices.com.au/sediment-trapping-field-guide/, accessed 3 March 2012).

NFF (2011) *Annual Review 2010–11*. National Farmers' Federation, Canberra.

SRDC (2012a) SRDC grower group innovation project wins national research award (http://www.srdc.gov.au/pages.aspx?id=20, accessed 3 March 2012).

SRDC (2012b) Resources for grower groups (http://www.srdc.gov.au/pages.aspx?id=31, accessed 3 March 2012).

SRDC (2012c) SRDC Grower Innovation Awards (http://www.srdc.gov.au/pages.aspx?id=43, accessed 3 March 2012).

9 Fostering Sustainable Farming

Ensuring farming is sustainable is important for any nation. This requires the farming community not only to make use of the natural environment in which they operate but also to protect it. Protecting the environment and natural resources, however, is not just the farmers' responsibility: it also needs the support of the general public and the effort of society as a whole. Environmental issues concerning the Australian farming community and Australia's initiatives in dealing with such issues are addressed in this chapter.

9.1 A Brief History of Environmental Issues in Australia

Prior to European settlement, native vegetation covered most of Australia. The Aboriginal people relied on the native fauna and flora. They lived with their environment in a harmonious way. It was not until European settlement that damage to the Australian natural environment commenced. Such damage has been brought about in two major ways: extensive land clearing and the introduction of fauna and flora by new settlers.

9.1.1 Land clearing

Land clearing in Australia is the removal of native vegetation and habitats, including the bulldozing of native bushlands, forests, savannah, woodlands and native grasslands, and the draining of natural wetlands for replacement with agriculture, urban and other land uses. Following European settlement, land clearing took place gradually as settlers moved into new areas. During early settlement, land clearing was a slow process as only some hand tools such as axes were available.

It was in the 1920s–1950s, when large-scale land clearing machinery was invented and used, that larger areas of land were cleared. Public policy encouraged land clearing as a way of expanding agricultural land. Prior to the 1980s, the general public had little concern with land clearing, as agricultural activities provided much wealth for the country. As much as 70% of Australia's native vegetation has been cleared or modified in the past 220 years, most of which occurred between the 1920s and 1970s. Extensive land clearing, however, has resulted in much, unanticipated, serious damage to the environment and has been detrimental to the lives of the Aboriginal people. One of the major problems caused by land clearing is soil salinity.

Salinity is the movement of salt to the land surface via groundwater. In Australia, there are vast amounts of salt stored beneath the land surface because of its geological history. Much of Australia's native vegetation has adapted to low rainfall conditions and uses deep root systems to take advantage of any available water beneath the surface. In this way, plant roots help to keep salt in the earth, by keeping groundwater levels low enough so that salt is not pushed to the surface. Prior to British settlement in 1788, groundwater levels were in equilibrium; the salt water of the underground tables rarely rose to ground level due to absorption by native flora growing on the surface. After the clearing of the native vegetation for European-style agriculture, excess water leached down into the saline layers of the soil and allowed the salt to move, sometimes into waterways or to the soil surface. Apart from dryland salinity, irrigation salinity also occurred in areas where irrigation was introduced and the excess loss of irrigation water below the root zone caused the water table to rise and bring up the salt.

In addition to the soil salinity problem, large-scale land clearing also results in other problems, such as soil erosion, loss of animal and plant species, and impact on climate.

Soil erosion

Land clearing exposes soil to erosion and leads to loss of soil and nutrient depletion. Vegetation removal results in less biological matter available for breakdown and replenishment of the nutrients in the soil.

Biodiversity

In Australia, the loss of some animal and plant species is attributed to land clearing. Vegetation removal can lead to the extinction of some native animals and plants due to the disturbance of their habitat. The health and resilience of the species that remain is also largely dependent on the size of the fragments of native vegetation and their distance from each other. The smaller and more isolated the remnants, the greater the threat to remaining species.

Impact on climate

Vegetation absorbs carbon dioxide. Land clearing reduces such capacity. Land clearing, though on a global scale but not necessarily at a regional scale, is also believed to be related to higher temperatures, decreased rainfall and more intense droughts. The removal of vegetation damages the microclimate by removing shade and reducing humidity.

9.1.2 Introduction of foreign fauna and flora

Following the settlement of Europeans, some overseas fauna and flora were introduced to Australia, helping the settlers to create a 'home country' atmosphere, or for various other purposes. For example, foxes and rabbits were introduced for hunting; rubber vine and prickly acacia as ornamental plants; the cane toad to fight the cane beetle; and dung beetles to control the polluting effects of cattle dung. Of those introduced animal species, cane toads, foxes and rabbits have caused major damage to the environment. Examples of introduced plant species that have caused major damage include mimosa (terrestrial weed) and salvinia (aquatic weed).

9.1.3 Earlier efforts in environment conservation

In the 1960s, some pioneer conservationists started to draw people's attention to environmental and sustainability issues. In August 1966, the Australian Conservation Foundation (ACF) was officially born, with a vision to inspire a society that was environmentally aware and responsible. In the 1970s, the general public became more interested in environment protection. Since then, the Australian public has developed stronger concerns for the environment. Environmental issues such as soil salinity, soil erosion, biodiversity, water quality and climate change have become prominent in public debates during the past few decades.

As many of the problems associated with land clearing began to manifest themselves in the landscape during the mid-1980s, it was no longer encouraged but has been constrained tightly through regulations. Broad-scale clearing for agriculture took place some 30–40 years later in QLD than in southern states. The emergence of problems associated with large-scale land clearing such as soil salinity and soil erosion, which generally took about 20–30 years to appear, served as a useful warning to QLD. QLD became aware of the problems associated with land clearing and hence refrained significantly from clearing the land, starting from the early 1990s. In 2009, QLD banned broad-scale tree clearing for agriculture (P. Elliot, Townsville, 18 July 2011, personal communication). Land clearing in many states of Australia today is strictly controlled.

None the less, problems caused by previous land clearing continue to burden society. In the meantime, other environment problems such as soil acidity remain significant. Damage caused to the environment by foreign animal and plant species continues to spread, such as chinee apple, rubber vine, parthenium, prickly acacia, foxes, rabbits, cane toads, European carp, tilapia, sparrow and myna birds. Among them, the fast spread of cane toads remains a major headache. To combat these problems, public expenditure has increased substantially since the 1980s (Pannell, 2011).

9.2 Environmental Issues Concerning the Farming Community

Major environment issues affecting the Australian farming community are summarized in a publication by the Australian Bureau of Statistics (ABS), *Natural*

Resource Management on Australian Farms 2006–07. This publication is based on the ABS Natural Resource Management Survey 2006–2007, which was conducted from September 2007. This is the second of an ongoing biennial collection of NRM data. Published in 2008, it is the latest issue available. The first such survey was conducted in 2005, collecting data for 2004–2005. The 2006–2007 survey essentially included all agricultural businesses above a minimum size cut-off (AUS$5000, Estimated Value of Agricultural Operations), recorded on the Australian Business Register (ABR) maintained by the Australian Taxation Office, which operated in Australia at any time during the year ended 30 June 2007.

The survey asked managers of agricultural businesses to identify the extent and type of natural resource issues that affected their operations, including weeds, pests and land and soil problems that were present on their land, and the activities they undertook to prevent or manage them. It also asked managers to provide details of the costs and effort spent on addressing these problems.

According to the 2006–2007 survey, weeds, pests and land- and soil-related problems are the major environmental issues concerning the farming community. In 2006–2007, 94.3% of Australian agricultural businesses reported undertaking NRM activities to prevent or manage weeds, pests and land and soil. The most commonly reported NRM problems were pests (70.4% of agricultural businesses), followed by weeds (66.0%) and land and soil (56.4%). Weed-related management activities were reported by 88.8% of agricultural businesses, followed by pest-related management activities (80.4%) and land- and soil-related management activities (60.1%). A higher proportion of farmers reported undertaking activities than those reporting a problem, indicating that NRM problems are managed preventatively as well as remedially (ABS, 2008, p. 7).

Managing NRM-related problems has cost farmers dearly. In total, in 2006–2007, undertaking NRM activities cost almost AUS$3bn, or AUS$21,094 per agricultural business or AUS$7522 for each 1000 ha under management (ABS, 2008, p. 6). Of this AUS$3bn NRM expenditure, AUS$1574m, AUS$768m and AUS$649m were spent on the management of weeds, pests and land and soil problems, respectively. A total of almost 9.4 million person days was spent addressing these problems, an average of 66 person days per agricultural business reporting NRM activities, or 24 person days/1000 ha under management.

Provided below is a snapshot of the weeds, pests and land and soil problems at the state and industry levels.

9.2.1 State level

Weeds

The most common weed-related problem reported was decreased value of production (76.1%), followed by decreased value of holding (34.3%) and increased fire risk (32.0%). In the NT, after decreased value of production (65.9%), increased fire risk was the second most commonly reported weed-related problem, with 52.8% of agricultural businesses reporting this problem. In TAS and WA, agricultural businesses also commonly reported increased fire risk as a weed-related problem (44.0% and 46.0%, respectively) (Table 9.1).[1]

Table 9.1. Weed-related problems and activities by state (based on ABS, 2008, p. 15).

	NSW	VIC	QLD	SA	WA	TAS	NT	Australia
Number								
Agricultural businesses	47,629	37,410	30,551	15,815	13,592	4,766	640	150,403
Agricultural businesses reporting weed-related problems	32,468	23,403	20,776	9,924	9,145	3,101	405	99,222
Agricultural businesses reporting weed-related activities	43,278	33,552	26,168	14,443	11,610	3,979	548	133,578
Weed-related problems – proportion reporting (%)								
Decreased value of production	78.5	74.2	74.9	77.9	73.9	73.9	65.0	76.1
Decreased value of holding	38.9	31.1	38.8	28.4	23.1	32.4	27.7	34.3
Increased fire risk	32.5	31.1	23.6	32.3	46.0	44.0	52.8	32.0
Decreased native plant or animal populations and distributions	29.2	20.3	27.5	19.2	21.3	21.1	29.0	24.7
Blocked watercourses	18.8	24.0	21.1	12.9	20.5	34.7	17.7	20.5
Poisoned stock	22.8	10.2	28.6	12.0	21.2	10.4	16.1	19.4
Other weed-related problems	44.3	43.0	41.0	39.6	38.0	43.9	35.0	42.2
Weed-related activities – proportion reporting (%)								
Application of herbicides	86.7	89.8	86.6	93.1	91.4	84.5	87.2	88.5
Pulling manual removal or chipping	60.3	53.2	54.3	51.9	52.1	55.8	39.9	55.5
Slashing, cutting or mowing	55.0	54.2	56.4	57.9	46.5	63.5	65.5	55.0
Crop or grazing management	41.2	36.9	31.1	47.3	53.1	32.6	17.8	39.5
Cultivation	33.9	30.1	43.2	39.9	30.6	32.0	13.3	35.0
Burning	14.6	12.7	23.7	14.1	24.8	20.4	24.5	17.0
Use of biological control	6.3	2.4	5.5	5.3	3.6	8.7	9.2	4.9
Other weed-related activities	1.9	2.1	2.9	1.6	1.9	2.9	4.1	2.2

Pests

The presence of feral and domestic animals was the most commonly reported pest type (76.7% of agricultural businesses reporting pests), followed by native animals and birds (69.4%), insect pests (61.9%) and other pests (including parasites, slugs, nematodes, mites, etc.) (44.5%) (Table 9.2).

Native animals and birds were the most commonly reported pest types in TAS (83.4% of agricultural businesses reporting any pest), QLD (78.5%) and the NT (68.8%), while feral and domestic animals were the most commonly reported pest type in the remaining states. Feral and domestic animals were reported as pests by 44.6% of agricultural businesses in the NT reporting pest-related problems, against a national average of 76.7%. Insect pests were reported as common pests in SA (71.9%) and WA (70.2%), compared to a national average of 61.9% and a low of 50.4% in TAS (Table 9.2).

Of agricultural businesses reporting pest-related problems, crop damage or decreased crop production was the most commonly reported problem (67.3%), followed by decreased livestock production (54.9%).

Land and soil

Of agricultural businesses reporting land- and soil-related problems, those reported most frequently were erosion (48.3%), soil compaction (43.3%) and soil acidity (42.0%). Dryland salinity was reported most commonly in WA (44.9%), against a national average of 17.4%, while irrigation salinity was reported most commonly in SA (17.6%), against a national average of 7.3% (Table 9.3).

While erosion was the most frequently reported land- and soil-related problem, the area affected was greatest for soil compaction (16.1 million ha (Mha) of agricultural land or 3.8% of agricultural land nationally). Soil acidity was the next most widespread land- and soil-related problem (13.5 Mha, 3.2% of agricultural land nationally) followed by erosion (12.4 Mha, 2.9% of agricultural land nationally).

At the state level, VIC reported the highest proportion of area affected by soil compaction (1.6 Mha, 12.5% of VIC agricultural land) and the highest reported area of soil acidity (2.2 Mha, 16.6% of VIC agricultural land). NSW (including the ACT) reported the highest proportion of area affected by erosion (3.3 Mha, 5.6% of NSW/ACT agricultural land).

Although reported to be affecting less than 0.6% of agricultural land nationally, dryland salinity was reported to be affecting 2.0% of agricultural land in VIC.

9.2.2 Industry level

At the industry level, of the agricultural businesses comprising sheep, beef cattle and grain farming (60.8% of agricultural businesses nationally), 65.7% reported land- and soil-related problems (56.5% average across all industries), 63.9% reported weed-related problems (66.0% average across all industries) and 63.6% reported pest-related problems (70.4% average across all industries). Sheep, beef cattle and grain farming has the largest number of agricultural businesses in its industry (60.8% out of the national total), followed by fruit and tree nut growing

Table 9.2. Pest-related problems and activities by state (based on ABS, 2008, p. 19).

	NSW	VIC	QLD	SA	WA	TAS	NT	Australia
Number								
Agricultural businesses	47,629	37,410	30,551	15,815	13,592	4,766	640	150,403
Agricultural businesses reporting type of pest	40,944	30,137	26,248	13,074	11,881	3,889	553	126,726
Agricultural businesses reporting pest-related problems	33,932	24,356	22,308	11,100	10,306	3,482	462	105,947
Agricultural businesses reporting pest-related activities	39,041	29,241	24,647	12,456	11,272	3,798	509	120,963
Type of pests – proportion reporting (%)								
Feral and domestic animals	80.6	81.0	67.1	78.0	78.2	61.5	44.6	76.7
Native animal and bird pests	73.4	54.3	78.5	62.2	76.4	83.4	68.8	69.4
Insect pests	61.4	58.6	59.6	71.9	70.2	50.4	62.2	61.9
Other pests	42.0	45.8	39.6	52.5	50.0	52.6	38.1	44.5
Pest-related problems – proportion reporting (%)								
Crop damage or decreased crop production	63.7	69.3	57.7	78.9	78.4	81.1	58.3	67.3
Decreased livestock production	60.4	52.4	55.2	48.0	53.6	43.3	43.4	54.9
Decreased native plant or animal populations and distribution	32.9	26.1	22.4	23.0	28.3	25.3	19.7	27.3
Other pest-related problems	18.7	17.0	16.4	17.3	21.5	20.0	19.5	18.0
Pest-related activities – proportion reporting (%)								
Use of pesticides and/or insecticides	83.3	83.1	78.9	87.1	85.7	74.3	75.8	82.6
Shooting/trapping	55.3	57.2	54.2	54.1	66.3	76.8	45.9	57.0
Baiting	40.4	22.2	34.9	41.5	34.2	10.8	37.8	33.5
Crutching	37.0	32.6	4.5	49.1	51.7	31.4	0.2	31.6
Fencing and/or netting	24.0	21.4	16.7	19.1	28.3	41.4	25.6	22.3
Crop or grazing management	20.1	19.3	17.7	23.0	23.9	23.0	9.4	20.1
Use of introduced biological control	7.7	5.0	7.9	8.1	5.7	8.3	16.3	7.0
Other pest-related activities	8.7	8.4	5.7	9.2	10.4	7.2	7.5	8.2

Table 9.3. Land- and soil-related problems and activities by state (based on ABS, 2008, p. 23).

	NSW	VIC	QLD	SA	WA	TAS	NT	Australia
Number								
Agricultural businesses	47,629	37,410	30,551	15,815	13,592	4,766	640	150,403
Agricultural businesses reporting land- and soil-related problems	28,027	21,203	15,272	8,673	9,104	2,364	279	84,922
Agricultural businesses reporting land- and soil-related activities	29,709	22,723	16,739	9,530	8,942	2,473	253	90,368
Land- and soil-related problems – proportion reporting (%)								
Erosion	52.5	37.4	59.4	43.7	46.1	47.8	59.4	48.3
Soil compaction	42.0	43.2	50.2	42.0	39.4	37.2	15.0	43.3
Soil acidity	47.9	48.0	17.7	30.4	60.4	49.1	13.6	42.0
Surface waterlogging	15.8	24.9	19.0	18.3	42.8	41.2	30.8	22.6
Dryland salinity	13.1	16.8	6.4	24.4	44.9	13.6	2.4	17.4
Soil sodicity	13.8	17.3	11.7	23.4	24.1	8.1	8.5	16.2
Irrigation salinity	4.2	9.7	5.6	17.6	4.7	6.0	4.3	7.3
Other land- and soil-related problems	8.4	8.6	8.4	8.9	6.4	11.3	9.3	8.4
Land- and soil-related activities – proportion reporting (%)								
Crop and/or pasture type or management	56.3	54.8	45.6	58.5	63.7	50.3	37.1	54.7
Addition of soil conditioners	47.8	60.5	39.4	45.7	60.9	62.4	52.8	50.9
Grazing management	48.6	41.2	42.0	45.1	49.9	50.8	33.5	45.3
Soil testing	37.6	40.4	27.2	32.4	52.1	44.8	37.6	37.5
Changed cultivation methods or practices	33.6	27.9	30.9	42.9	34.9	23.7	12.7	32.4
Tree or shrub planting or maintenance	34.9	33.6	14.7	27.7	39.6	30.1	16.9	30.4
Construction of earthworks	31.6	24.5	38.5	14.2	30.8	30.5	42.6	29.2
Construction or maintenance of fencing	29.0	29.3	19.9	23.2	34.4	37.8	18.0	27.5
Changed fertilization methods or practices	23.2	25.3	17.0	20.1	25.9	27.4	17.9	22.6
Changed irrigation methods or practices	8.3	10.4	9.2	14.0	4.0	15.9	6.4	9.4
Other land- and soil-related activities	6.8	5.6	10.7	6.6	5.7	8.9	14.3	7.1

(9.0%) and dairy cattle farming (5.9%). The sheep, beef cattle and grain farming industry had the highest proportion of holdings reporting NRM-related problems, followed by other crop growing industries (Table 9.4). Poultry farming had the smallest proportion of holdings reporting NRM-related problems.

As noted earlier, NRM-related problems have been costly for farmers to deal with. They reduce production and cost money to fix or prevent. Farmers have made enormous efforts to deal with these problems. Their efforts alone, however, are not enough to reduce or eliminate such problems. Concerted efforts are required from governments, industries, the farmers themselves and broad communities. In the next section, we highlight the various initiatives and approaches developed in Australia to handle environment-related problems to support sustainable agriculture.

9.3 Initiatives and Approaches in Dealing with Environmental Problems

Australian society has become much more environment conscious over the past two decades as a result of proactive and widespread education. Both public and private agencies have played a significant role in educating society to care for the environment. The Australian public has generally accepted that our environment must be protected and many individuals participate voluntarily in caring for the environment in one way or the other. Children are taught to protect the environment through school education and various public media. The desire of the public to care for the environment in turn exerts pressure on farmers to adopt farming practices that are more environment friendly and on the governments to have legislations and acts to guide the farming community to practise sustainable farming.

Australian governments have been very responsive to demands to care for the environment. Federal, state and local governments all have a department or section that is designated to look after any environment-related issues. Environment legislations and acts have been developed and updated by various levels of governments to guide efforts in dealing with environment problems peculiar to their jurisdiction. Given that over 50% of Australia's landscapes are for agricultural production, much of the government environment effort is related to fostering sustainable farming.

Efforts to promote, help or practise sustainable farming occur at various levels: government (federal, state and local), region, industry, community and farm. The rest of this section provides examples of the initiatives and approaches developed and used for sustainable farming at different levels.

9.3.1 Federal government

Through the Department of Sustainability, Environment, Water, Population and Communities and DAFF, the Australian Government delivers a number of environment and sustainable agriculture programmes. One of the current major programmes is 'Caring for our Country' (CFC).

Table 9.4. NRM-related problems and activities by industry (based on ABS, 2008, p. 30).

		Agricultural businesses reporting NRM-related problems			Agricultural businesses reporting NRM-related activities		
	Agricultural businesses	Weed-related problems	Pest-related problems	Land- and soil-related problems	Weed-related activities	Pest-related activities	Land- and soil-related activities
011 Nursery and floriculture production							
0111 Nursery production (under cover)	421	191	363	97	345	371	123
0112 Nursery production (outdoors)	780	425	579	265	750	625	345
0113 Turf growing	301	185	162	164	271	197	174
0114 Floriculture production (under cover)	258	99	168	68	208	199	108
0115 Floriculture production (outdoors)	479	342	374	222	446	379	206
Subtotal	2,239	1,242	1,646	816	2,020	1,771	956
Proportion reporting (%)		***55***	***74***	***36***	***90***	***79***	***43***
012 Mushroom and vegetable growing							
0121 Mushroom growing	97	[a]	47	[a]	51	61	[a]
0122 Vegetable growing (under cover)	913	493	530	329	710	600	392
0123 Vegetable growing (outdoors)	4,055	2,528	2,890	2,026	3,584	3,224	2,467
Subtotal	5,065	3,021	3,467	2,355	4,345	3,885	2,859

Proportion reporting (%)		***60***	***68***	***46***	***86***	***77***	***56***
013 Fruit and tree nut growing							
0131 Grape growing	6,039	3,012	3,281	2,696	5,553	4,012	3,188
0132 Kiwifruit growing	46	44	a	41	a	a	27
0133 Berry fruit growing	465	275	361	267	440	380	272
0134 Apple and pear growing	780	539	684	341	742	765	456
0135 Stone fruit growing	1,193	704	878	604	1,134	1,051	641
0136 Citrus fruit growing	1,388	769	992	712	1,254	1,141	703
0137 Olive growing	398	180	262	215	369	285	224
0139 Other fruit and tree nut growing	3,271	2,147	2,599	1,978	3,077	2,654	2,071
Subtotal	13,580	7,670	9,057	6,854	12,569	10,288	7,582
Proportion reporting (%)		***56***	***67***	***50***	***93***	***76***	***56***
014 Sheep, beef cattle and grain farming							
0141 Sheep farming (specialized)	12,150	8,447	9,932	7,889	10,468	11,069	7,899
0142 Beef cattle farming (specialized)	44,957	30,547	30,371	24,597	39,232	35,337	24,822
0143 Beef cattle feedlots (specialized)	243	151	158	148	191	167	160
0144 Sheep–beef cattle farming	8,501	5,808	7,004	5,577	7,446	7,887	5,544
0145 Grain–sheep or grain–beef cattle farming	14,131	10,318	11,485	9,707	12,985	12,706	10,911
0146 Rice growing	127	74	81	77	127	109	100

(*Continued*)

Table 9.4. *(Continued)*

		Agricultural businesses reporting NRM-related problems			Agricultural businesses reporting NRM-related activities		
	Agricultural businesses	Weed-related problems	Pest-related problems	Land- and soil-related problems	Weed-related activities	Pest-related activities	Land- and soil-related activities
0149 Other grain growing	11,358	8,070	8,387	7,774	10,543	9,355	8,252
Subtotal	91,467	63,415	67,418	55,769	80,992	76,630	57,688
Proportion reporting (%)		***69***	***74***	***61***	***89***	***84***	***63***
015 Other crop growing							
0151 Sugarcane growing	3,975	2,891	2,835	2,334	3,634	2,870	2,607
0152 Cotton growing	526	317	418	314	507	467	397
0159 Other crop growing n.e.c.	2,472	1,536	1,628	1,193	2,163	1,651	1,281
Subtotal	6,973	4,744	4,881	3,841	6,304	4,988	4,285
Proportion reporting (%)		***68***	***70***	***55***	***90***	***72***	***61***
016 Dairy cattle farming							
0160 Dairy cattle farming	8,921	5,503	5,708	4,642	7,964	6,956	5,384
Proportion reporting (%)		***62***	***64***	***52***	***89***	***78***	***60***
017 Poultry farming							
0171 Poultry farming (meat)	797	350	302	190	595	541	249
0172 Poultry farming (eggs)	499	217	209	139	392	323	118
Subtotal	1,296	567	511	329	987	864	367

Proportion reporting (%)		***44***	***39***	***25***	***76***	***67***	***28***
018 Deer farming							
0180 Deer farming	184	83	118	103	152	141	101
Proportion reporting (%)		***45***	***64***	***56***	***83***	***77***	***55***
019 Other livestock farming							
0191 Horse farming	3,019	1,947	1,456	1,314	2,706	2,048	1,671
0192 Pig farming	894	481	559	431	768	755	421
0193 Beekeeping	30	[a]	[a]	[a]	[a]	[a]	[a]
0199 Other livestock farming n.e.c.	1,049	599	632	428	734	566	464
Subtotal	4,992	3,027	2,647	2,173	4,208	3,369	2,556
Proportion reporting (%)		***61***	***53***	***44***	***84***	***67***	***51***
All agriculture	134,718	89,304	95,499	76,906	119,598	108,943	81,806
Proportion reporting (%)		***66***	***71***	***57***	***89***	***81***	***61***
All other industries	15,685	9,918	10,448	8,016	13,981	12,020	8,563
Proportion reporting (%)		***63***	***67***	***51***	***89***	***77***	***55***
Total all industries	150,403	99,222	105,947	84,922	133,578	120,963	90,368
Proportion reporting (%)		***66.0***	***70.4***	***56.5***	***88.8***	***80.4***	***60.1***

Notes: [a]Not available for publication but included in totals where applicable; some numbers are estimates; n.e.c. = not elsewhere classified.

CFC, a Commonwealth government initiative, seeks to achieve an environment that is healthy, better protected, well managed and resilient, and which provides essential ecosystem services in a changing climate. An overview of the programme including its six priority areas, outcomes and business planning and targets is shown in the diagram below (Fig. 9.1).

Through this programme, the Australian Government is investing AUS$2bn in NRM over a 5-year period (2008–2013). The programme funds projects across the country to achieve national targets – projects that improve biodiversity and sustainable farm practices. This funding supports regional NRM groups, local, state and territory governments, indigenous groups, industry bodies, land managers, farmers, Landcare groups and communities.

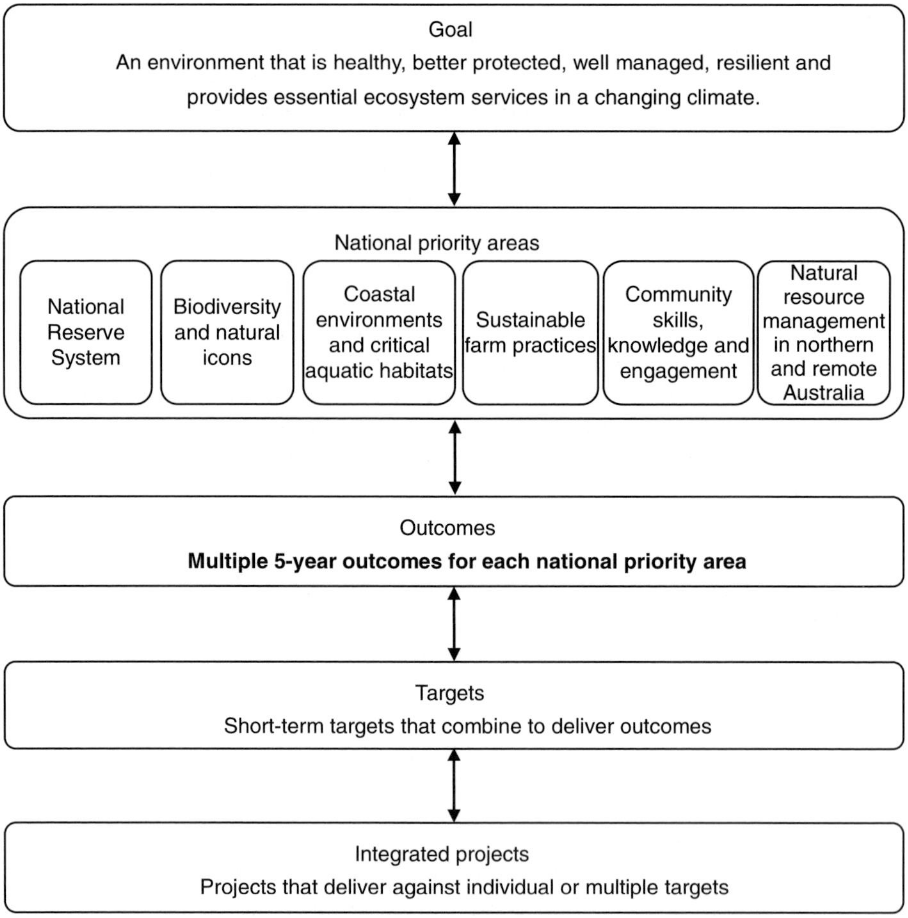

Fig. 9.1. CFC goal and priority areas (Australian Government Land and Coasts, 2011).

As shown in the diagram, CFC is currently investing funds to improve strategic outcomes across six national priority areas, namely:

- the National Reserve System;
- biodiversity and natural icons;
- coastal environments and critical aquatic habitats;
- sustainable farm practices;
- community skills, knowledge and engagement; and
- natural resource management in northern and remote Australia.

All the six priority areas contribute to improving the environment in which the farming sector operates. The area of 'sustainable farm practices' is devoted especially to improving the environmental outcomes from farming. By the end of the first 5 years of CFC, that is, 2013, the 'sustainable farm practices' programme aims to: (i) assist at least 30% of farmers to increase their uptake of farm and land management practices that deliver improved ecosystem services; (ii) increase the number of farmers who adopt stewardship, covenanting, property management plans or other arrangements to improve the environment both on farm and off farm; and (iii) improve the knowledge, skills and engagement of at least 30% of land managers and farmers in managing natural resources and the environment.

Investments under the CFC programme also complement other Australian Government programmes such as Water for the Future and Australia's Farming Future. CFC integrates the Australian Government's previous NRM initiatives, including the Natural Heritage Trust, the National Landcare Program, the Environmental Stewardship programme and the Working on Country Indigenous land and sea ranger programmes. Further information about the monitoring, evaluation, reporting and performance improvement strategy can be obtained from the official site of the CFC programme at: http://www.nrm.gov.au/me/index.html.

Another priority area the Australian Government deals with is climate change. Climate change has been regarded increasingly as a major threat to the health of the Australian environment. According to CSIRO research (Preston and Jones, 2006; CSIRO, 2007) and the Garnaut Climate Change Review and its update (Garnaut, 2008, 2011), climate change is expected to have numerous adverse effects on many species, regions, activities and much infrastructure and areas of the economy and public health in Australia.

The Australian Government recognizes climate change is occurring and has been putting strategies in place either to mitigate or to reduce the impact on regions and people. One important strategy is related to cutting greenhouse gas emissions. Australia has one of the highest rates of CO_2 emissions from energy use per capita (15.8 t in 1995), according to OECD estimates (ABS, 2000, p. 506). Only Canada, the USA and Luxembourg experience higher levels. Hence, gas emissions need to be reduced.

One of Australia's first national attempts to reduce gas emissions was the voluntary-based initiative called the Greenhouse Challenge Program, which began in 1995 (Parliament of Australia, 2003). In November 1997, the then Prime Minister John Howard announced a package of other measures designed to address climate change, including measures that focused on reducing the environmental impacts of the energy sector – *Safeguarding Our Future: Australia's Response to Climate Change* (ABS, 2000, p. 506). One measure was the establishment of the Australian Greenhouse Office (AGO) in 1998, which was then the world's first government agency dedicated to cutting greenhouse gas emissions (AGO, 2011).

After contributing to the development of, then signing but not ratifying the Kyoto Protocol, action to address climate change was coordinated through the AGO. The AGO released the *National Greenhouse Strategy* in 1998. The report recognized climate change was of global significance and that Australia had an international obligation to address the problem. In 2000, the Senate Environment, Communications, Information Technology and the Arts References Committee (2000) conducted an inquiry that produced *The Heat is On: Australia's Greenhouse Future*.

Climate change featured strongly in the November 2007 Australian federal election in which John Howard was replaced by Kevin Rudd as prime minister. The first official act of the new Australian Government was to ratify the Kyoto Protocol. Under the Rudd government, the new Department of Climate Change was established to coordinate and lead climate policy in the Australian Government and aimed to have a national emissions trading scheme operating by 2010. However, on 27 April 2010, Prime Minister Kevin Rudd announced that the government had decided to delay the implementation of the Carbon Pollution Reduction Scheme (CPRS) until the end of the first commitment period of the Kyoto Protocol (ending in 2012) (Department of the Prime Minister and Cabinet, 2010). The government cited the lack of bipartisan support for the CPRS and slow international progress on climate action as the reasons for the decision.

In February 2011, Prime Minister Julia Gillard announced a plan to implement a fixed price to be imposed on carbon pollution from 1 July 2012. The carbon tax would operate for 3 to 5 years before a full emissions trading scheme was implemented. On Sunday, 10 July 2011, Julia Gillard announced a carbon tax package. The price was to be AUS\$23/t, moving to a floating emissions trade scheme in 3 years. The revenue is expected to be about AUS\$10bn annually. Part of the money will be used for household compensation (to eligible income groups), assistance to high-emitting industries to upgrade technology and reduce emissions, renewable energy R&D and soil carbon sequestration.

Reduced carbon emissions, and hopefully the resultant less drastic climate changes, will help farm production. However, before farmers can potentially benefit, they have a carbon price to pay, though not directly. The tax on carbon will add to farmers' input costs because fuel, electricity, fertilizer and farm chemicals that are petroleum- or coal-based will rise in price. To encourage farmers to participate in combating carbon emission, they are provided incentives for soil carbon sequestration and for building skills to meet climate change. The Australian Government is to buy credits from farmers who make carbon savings (e.g. through planting trees) or to allow them to sell these credits to polluters to cover their carbon bill. The government will also reinvest carbon tax revenue in land research and management programmes, including funding for outreach officers and training for farmers who want to take part in the scheme.

9.3.2 State governments

State and territory governments are generally very proactive in developing legislation and guidance to require farmers in their states to protect their environment. Two examples of state-level initiatives are provided. One is vegetation management in QLD under the Vegetation Management Act 1999 and the other is pastoral management in SA by the Pastoral Board of South Australia. In other states and territories, similar arrangements exist; for example, the Native Vegetation Regulation 2005, which underpins the Native Vegetation Act 2003 in NSW, and the Pastoral Lands Board of Western Australia in WA.

Vegetation management in Queensland

An adequate level of vegetation plays an important role in keeping the environment healthy. As has been pointed out earlier, large-scale land clearing, and thus the disappearance of much native vegetation in Australia, has caused serious land degradation and damage to natural resources. QLD's large-scale land clearing took place some 30–40 years behind those states in the south, enabling them to learn lessons from the southern states. To manage the state's native vegetation better, a Vegetation Management Act was issued in 1999. The Vegetation Management Act 1999 was established to help the state to manage its remnant woody vegetation. The state's vegetation management is administered by the Department of Environment and Resource Management (DERM).[2]

DERM is the agent of the QLD government that is entrusted to conserve, protect and manage the state's environment and natural resources for today and for future generations. It administers NRM and environmental protection legislation on behalf of the QLD government, and works collaboratively with government, business, industry and communities to conserve natural values, improve environmental performance, reduce pollution and encourage sustainable NRM behaviours (DERM, 2011). DERM has five key responsibilities, as described briefly below; this brief description helps to show where vegetation management sits and to appreciate the broad scope of environment and resource management efforts in QLD.

1. Leading environmental recovery: coordination of efforts for environmental recovery in response to natural disasters.
2. Meeting the challenge of climate change: responsible for developing initiatives which ensure QLD reduces its carbon footprint while also making sure the state is well placed to meet the impacts of climate change.
3. Conserving and enhancing the natural environment, ecosystems and cultural heritage: dealing with a wide range of environmental matters including protecting air, water and soil quality, managing waste, vegetation management, preventing or controlling pollution, managing the state's coastline and promoting sustainable industry.
4. Securing water for QLD's future: developing water resource plans for catchment areas throughout the state to provide a secure and reliable system for supplying water to QLD communities.

5. Managing QLD's land: managing and promoting the sustainable use of the state's land resources through the development of policies and programmes that are designed to ensure healthy landscapes.

Hence, vegetation management is part of DERM's third responsibility area: conserving and enhancing the natural environment, ecosystems and cultural heritage. In QLD, the vegetation management framework is composed of legislation, a state policy, regional vegetation management codes, an offsets policy, a regrowth vegetation code and material change of use and reconfiguring a lot policies.

LEGISLATION Vegetation management is regulated through the Vegetation Management Act 1999 and the Sustainable Planning Act 2009.

STATE POLICY The State Policy for Vegetation Management documents the QLD government's policy on vegetation management – defining the principles that underlie the policy, its desired outcomes and how these are achieved. The policy aims to ensure:

- conservation of biodiversity;
- maintenance of ecological processes;
- clearing does not cause land degradation;
- management of the environmental effects of clearing;
- reduction of greenhouse gases;
- balanced decision making; and
- support for regional communities.

REGIONAL VEGETATION MANAGEMENT CODES DERM assesses applications to clear vegetation against regional vegetation management codes. Four codes based on bioregional areas cover all areas of QLD.[3]

OFFSETS POLICY The offsets policy applies to an offset proposed to meet a performance requirement under a regional vegetation management code.

REGROWTH VEGETATION CODE New regrowth laws took effect on 8 October 2009. Clearing regulated regrowth must comply with the regrowth vegetation code, unless clearing is for an exempt activity.

MATERIAL CHANGE OF USE/RECONFIGURING A LOT POLICIES The department may be required, as a 'concurrence' agency, to assess any remnant vegetation clearing associated with a proposed material change of use (MCU) or a reconfiguration of a lot (RaL) application. The MCU and RaL policies are used to undertake this assessment.

Under this vegetation management framework, any intended clearing of native vegetation in QLD is now closely monitored, controlled and regulated. The *Vegetation Management Act 1999* sets down all the rules and regulations that guide *what* clearing can be done and *how* it must be done to meet the requirements of the law. Any violation of these rules and regulations is dealt with seriously. Subsequently, further damage to native woody vegetation in QLD is likely to be limited in the near future.

Pastoral Board of South Australia

More than 75% of Australia is defined broadly as rangelands (ANRA, 2009). Rangelands include a diverse group of relatively undisturbed ecosystems such as tropical savannahs, woodlands, shrublands and grasslands. Rangelands extend across low and variable rainfall climates, including arid, semi-arid and some seasonally high rainfall areas. Extensive grazing on native pastures occurs across the rangelands, while broad-scale cropping and cultivation generally do not take place.

In SA, the rangelands occupy some 741,000 km^2, encompassing a little over 80% of the state's land area. Within this area, hundreds of pastoral leases occupy collectively some 410,000 km^2. The balance of the rangelands is set aside as Aboriginal lands and parks and reserves (PBSA, 2010). These rangelands, however, are in the arid part of the state, with very low but erratic rainfall (annual average between 100 and 250 mm).

To ensure sustainable pastoral land use, the Pastoral Board of South Australia (PBSA) was established to help administer the *Pastoral Act of* 1893. It is one of the oldest continually active statutory authorities in Australia. The first board was given the power to determine lease areas, boundaries, rents and the duration of tenured occupancies. In the 1930s, the sustainability of the current grazing practices began to be questioned widely and openly. Short-term tenures, poor stocking practices and a run of bad seasons had all contributed to the dissatisfaction which gave rise to these questions. The *Pastoral Act of 1936* introduced stocking controls on leases for the first time.

Today, the sustainable management of pastoral lands subject to grazing, and the monitoring of the condition of these lands, are pivotal components of the work of the PBSA. The board also has a legally delegated responsibility from the Native Vegetation Council to administer the *Native Vegetation Act 1991* with respect to the clearance of native vegetation by grazing on pastoral leases. The status of the PBSA, its functions, its legislative objectives and details on various other operational aspects of the board can be found in its annual reports. Donovan (1995) presents an interesting reading on the history of the PBSA. He provides a historical perspective from the viewpoint of the administrators who have been involved in the development and management of SA's rangelands.

9.3.3 Local governments

Local governments also have various programmes that are related to environment protection and management. The City of Townsville, a coastal city in north QLD with a population of about 200,000, has a number of sustainability programmes including, for example, shoreline erosion management and climate change. For shoreline erosion management, two Shoreline Erosion Management Plans (SEMPs) have been developed. The plans provide a long-term, sustainable, coastal management approach to foreshore protection from erosion, while maintaining culture, heritage and environmental values. Environment education is also part of the sustainability programme.

The City of Orange is a smaller rural centre of some 40,000 people in the NSW Central West Tablelands, about 260 km west of Sydney. Surrounding the city are many farming operations ranging from horticulture, viticulture, grazing and broadacre farming. The city council follows the principles of ecologically sustainable development. It strives to use, conserve and enhance the community's resources so that ecological processes are maintained and the total quality of life, now and in the future, can be increased. Its Environmental Sustainability Action Plan provides direction on management issues associated with water, waste, energy, biodiversity, pollution and salinity. These issues are addressed with a range of actions that include education (and behavioural change) and responses to climate change.

9.3.4 Regional

Collaboration is often needed to coordinate environment protection and management efforts beyond a city or even a state. Cooperation between regions has been common in Australia. The above-mentioned Environmental Sustainability Action Plan in Orange is such an example. The three cities of Bathurst, Orange and Dubbo are next to each other, all in an agriculture-rich area of NSW. Their councils are committed to persistent progress towards sustainability, and to being leaders in delivering economic, environmental and social well-being. To this end, they developed a Memorandum of Understanding to promote skills and knowledge sharing and improved outcomes for their communities through a strategic alliance.

A priority for the alliance was to develop and implement an Environmental Sustainability Action Plan for all three cities. Orange City Council's Environment and Natural Resources Management Advisory Committee played a major role in ensuring local community needs were considered in developing and implementing actions of the plan. Collaboration and cooperation go beyond the three cities. The funds for developing the plan were provided primarily by the NSW Environmental Trust under the City and Country Environmental Restoration Program. The Central West Catchment Management Authority, NetWaste and the Central Regional Organization of Councils also made significant contributions to the plan's development. The joint efforts helped to ensure that the plan provided the necessary integration required to maximize environmental outcomes for the region.

Another example of regional collaboration is the 'NQ Dry Tropics NRM'. It is one of the 14 NRM groups under the 'Queensland Regional NRM Groups Collective'. The NQ Dry Tropics NRM is a community-based, not-for-profit organization established in 2002 to deliver on-ground NRM activities and enhance the community's involvement in such activities throughout the Burdekin region, which is located in north-eastern QLD. It covers an area of approximately 138,000 km^2. The NQ Dry Tropics NRM is defined primarily by the catchment area of the Burdekin River, reflecting a catchment-based approach that is popular among regional NRM groups. The region consists of six distinct bioregions.

The Burdekin region extends into marine waters and includes Magnetic Island and the Palm Islands. It covers all or part of 12 local government jurisdictions, including Townsville, Mackay, Charters Towers and Burdekin. The economy of the region is heavily reliant on natural resource-based industries, particularly agriculture and mining. Agriculture is by far the most important employer in the rural areas of the region. Other major industries include mining and tourism. Protecting the natural environment to ensure sustainable farming is therefore very important.

9.3.5 Industry

As noted earlier, in Australian agriculture, a large quantity of chemicals are used to control weeds and pests. Not only is their use costly, the containers of such chemicals and leftover chemicals can, if not disposed of properly, also cause serious damage to the environment and the health of livestock animals and humans. Adequate disposal of such wastes is hence very important. Those who are involved in the manufacturing, distributing and use of such chemicals got together with agricultural industry bodies, local councils and waste management and recycling companies and developed two programmes, drumMUSTER and ChemClear, for the safe disposal of such wastes. Participation in the programmes is voluntary. drumMUSTER collects used chemical containers. ChemClear focuses on removing unwanted crop protection and animal health chemicals from farms. Both programmes are under the supervision of AgStewardship Australia (www.agstewardshipaustralia.org.au). The establishment, operation and achievement of one of the two programmes, drumMUSTER, is highlighted below, which is based on the presentation by the CEO of AgStewardship Australia, Karen Gomez, at the 2011 ABARES Outlook Conference (Gomez, 2011).

In 1998, a national strategy for the management of agvet chemicals was put in place. In response, representatives of agricultural chemical manufacturers, distributors and end users got together to manage agvet chemicals themselves on a voluntary basis. Consequently, drumMUSTER was established and began operating in 1999.

A levy of 4 cents/l is collected from participating manufacturers, who then pass the levy on to the end users. The levy is used to fund the operations of the drumMUSTER programme, which is run on a daily basis. Farmers who participate in the programme buy chemicals that are produced by participating manufacturers, with a drumMUSTER logo on the container. After use, these farmers enjoy the convenience of disposing of the chemical containers to a local collection agency or centre. drumMUSTER has a network that collects these returned chemical drums, which are then recycled. Disposing of containers that do not have a drumMUSTER logo may end up costing more. This has encouraged most farmers to participate in the programme voluntarily.

To date, drumMUSTER has reduced 75% of the wastes that would otherwise have gone to landfill, which is translated to some 17 million drums since it started in 1999. Over 80 manufacturers and suppliers participate in the programme. The programme now covers some 90% of agricultural sales in Australia. It has over 700 collection sites across Australia; most of which are based in regional and rural local government waste transfer centres. Some 3000 trained inspectors work in the collection network to make sure that all returned drums are cleaned properly. Since 1999, drumMUSTER has developed high levels of brand awareness, diverse partner relationships and a substantial operational footprint. The programme has provided substantial benefits to the Australian public, environment and future generations. Farmers are generally very appreciative of being offered such a convenient and safe way to dispose of chemical drums.

Australia has a relatively large livestock sector. The greenhouse gas emission of this sector is also relatively large and is the main source of agriculture emissions. In 2009, agriculture produced an estimated 84.7 Mt CO_2 emissions. Of this, methane alone accounted for 54.7 Mt (or 64.6%). In other words, methane emission alone accounted for 10% of net national emissions (Department of Climate Change and Energy Efficiency, 2011, p. 203). Hence, reducing methane emissions is important for the livestock industry. It helps the industry to project an image in the minds of the public as being proactive in contributing to a better environment, thus winning over their support as consumers and also as supporters of the industry.

Recognizing it has a responsibility to manage the environment sustainably, the livestock industry, under the leadership of MLA, has invested significantly in R&D to develop ways to minimize its impact on the environment. Currently, MLA is coordinating and jointly funding a AUS$28m programme on behalf of, and in partnership with, the Australian Government. The overall objective of the research programme is to develop practical on-farm options to achieve a significant reduction in methane emissions from livestock, and to quantify the level of abatement achievable while at the same time increasing productivity. There are 18 projects grouped into six themes in the programme. The research programme focuses on reducing the amount of methane produced by ruminant animals in three key ways:

- manipulation of nutrition and rumen function to lower emissions;
- animal breeding to select animals that produce lower emissions; and
- whole-farm system approaches and better herd management.

Since 1990, emissions from producing beef in Australia have decreased by 6.5% per kilogram of beef produced; the only production industry in Australia that has reduced greenhouse gas emissions (MLA, 2010). This is attributable to improved productivity: the age to market of animals is decreasing while at the same time the amount of meat being produced from each animal, on average, is increasing due to more efficient production methods. Australian meat is produced with one of the lowest carbon emission profiles of any major meat-producing country in the world.

9.3.6 Community

At the community level, an important initiative by Australians in NRM is the grass-roots Landcare movement. This movement harnesses individuals and groups under the ethic of caring for the land. It had its genesis in initiatives to improve agricultural productivity through sustainable land management. The movement has grown from this to a broader focus on sustainable management of all of Australia's natural resource assets and now encompasses individuals and groups across the whole landscape from coastal to urban and remote areas of Australia (Landcare Australia, 2011).

A Landcare group usually starts when community members with common objectives connect over their observations of a local environmental issue. Groups set their own agenda, undertake work as often as they like and choose their own project sites. Groups may apply for funding from a variety of different sources, such as CFC, to support their work.

Generally, small group committees oversee operations, apply for project funding and organize communal activities like community workshops or tree planting. Most groups have one to six formal meetings annually. They may run discussion sessions, and short trips to other Landcare groups and other activities to gain and share knowledge. Some larger groups may have a paid coordinator providing part-time assistance, arranging meetings and activities and providing management guidance. Funds to pay these salaries come mostly from the governments. Increasingly, Landcare groups amalgamate into Landcare networks managed by community boards that take a more regional approach to environmental issues and coordinate activities to achieve catchment-wide outcomes. Networks are now a major community link to all levels of government and industry for financial support and information (Landcare Australia, 2011).

The first official Landcare group to form in Australia was the Winjallok Landcare group, near St Arnaud in VIC, on 25 November 1986 (Victorian Department of Primary Industries, 2011). In 1989, the Landcare movement officially began and Landcare became a national programme. The Australian Government, with bipartisan support, announced its 'Decade of Landcare Plan' and committed AUS$320m to fund the National Landcare Program (Landcare Australia, 2011). Today, there are more than 4000 community Landcare groups, 2000 Coastcare groups and many thousands of Landcare volunteers in most towns across the country. Two examples of Landcare groups follow.

Roper River Landcare Group, NT

About half of the NT's land is held under pastoral tenure for grazing cattle. Although some introduced pasture species have been cultivated since the 1930s, many of the pastoral properties in the NT rely on unimproved native pastures. Managing invasive species (including some pasture grasses) is also a concern shared by all land managers. The Roper River Landcare Group was established in 1993, initially in response to the noxious weed, Parkinsonia. Some of the key issues for the group today include feral animal control, tourism impacts and water quality, maintaining barramundi stocks, reducing litter and controlling erosion, getting local children involved in the Landcare movement and promoting the group and its activities.

The group promotes the adoption of best practice NRM management in order to achieve biodiversity and sustainable land use outcomes across the Roper River catchment. It aims to maintain and expand networks and NRM partnerships with catchment stakeholders, including indigenous organizations, pastoralists, horticulturalists, townspeople, other NRM groups and all levels of government, and use networks to build on stakeholder awareness of key NRM issues (weeds, threatened species, ferals, fire) and deliver skill- and capacity-building activities. Recently, the group successfully obtained a grant of over AUS$1.3m from the CFC programme for a project entitled 'Building capacity to protect the cultural, conservation and production values of Mangarrayi Traditional Lands', aimed at improving land management practices (Caring for our Country, 2011).

Little River Landcare Group, NSW[4]

The Little River Landcare Group is a Landcare network. It was started by a handful of dedicated people in the Little River catchment in late 1997. The Little River catchment is part of the Macquarie River valley in central western NSW. The Little River drains 258,000 ha from over 300 farms and centres on the small townships of Yeoval and Cumnock. Before the Little River Landcare Group was formed, there were already 14 individual Landcare groups within the catchment. The Little River Landcare Group united the 14 individual Landcare groups in order to make a concerted effort to improve the natural environment in their catchment. It represents an innovative approach to NRM within a catchment area. Its objective is to promote a healthy, productive and diverse biological and social environment.

To achieve its objective, the group realized that they needed a more sophisticated and wide-reaching approach than just planting trees, as many other Landcare groups usually do. It adopted a corporate business model, engaging staff and involving the community so as to harness best any available human capital. The group has a key focus to help people to change the way they think, as well as the way farmers manage their farms on a daily basis. It promotes this attitude towards handling environment problems: 'it's not your problem or my problem, it's our problem'.

The focus on engaging the wider community has helped the group to achieve impressive results. Its membership expanded from 30 to 300 in a few years. By 2010, it had 515 individual members and 200 business members and covered an area of 340,000 ha. The budget, too, has grown from AUS$20,000 to AUS$1m. This has allowed the group to expand its reach into many other activities such as education and field days. A programme called 'Positive Farming Footprint' has delivered basic extension skills to some members. Another programme titled 'Farmers Teaching Farmers' is intended to enable members to share their accumulated knowledge and experience.

Such activities have benefited members and the environment as well. The intensive and costly annual pasture and fertilizer regime of the past has been replaced by perennial pastures like native grasses and legumes, and these are grazed on a rotational basis. Conservation farming practices such as minimum tillage help members to protect their most valuable resource – soil. Water quality has improved as a result of more on-ground cover and less runoff. In short, the integrated approach adopted by the Little River Landcare Group has enabled concerted efforts by its members, which in turn have resulted in major long-term improvements in the health of the Little River catchment.

The strong Landcare presence in Australia has encouraged diverse community activities addressing environmental, economic and social issues. Today, more than 40% of farmers are involved in Landcare activities, and many more practice Landcare farming. They make significant contributions to combating soil salinity and erosion through sound land management practices. Landcare groups and networks will continue to provide the cornerstone of the unique government–community partnership for Australia's NRM.

9.3.7 On farm

No one understands the importance of land to one's livelihood better than farmers. Generally, farmers anywhere in the world have a desire to maintain good stewardship of the land and to pass on their land to their heirs in as good or better condition than it came to them. To say farmers do not care about their land is not fair and is flawed. Of course, there are farmers who exploit the land for short-term gains, but more often because they are victims of circumstances over which they may have no control, such as very small holdings, low product prices or adverse seasons. To characterize the majority of farmers as deliberate exploiters is false and malicious (Campbell, 1980, p. 230).

In Australia, the majority of farmers are very conscious in caring for their land. During 2010–2012, the author had the opportunity to visit a number of farmers in various states. All of them understood well the importance of natural resources to their livelihood and carried out activities to look after their land. A few examples follow.

Wes Brown, manager at 'Grenabri' near Cargo in the Central West Tablelands of NSW has developed his own resource management philosophy. He believes 'no pasture, no livestock production'. He continuously explores ways for sustainable use of the pasture on the farm he manages. One approach he has developed is to subdivide paddocks further into smaller fenced blocks, coupled with careful selection of a grass combination suitable for the blocks. Although this approach requires him to move the animals between blocks more frequently, it helps avoid overgrazing, thus protecting the feed base and reducing runoff.

Ross and Prue Leckie live and farm on a property, 'Rocky Hill', in Charleston, a small country town that is about 25 km east of Adelaide in SA. They are very much dedicated to the protection of their environment and are actively involved in the activities of their local Landcare group, New Springs Landcare Group. As farmers, they understand well the importance of a sound environment to sustainable farming. They care about the land on their farm and have continuously made improvements to their paddocks. They have even fenced off one area of their farm just for conservation purposes (Ross Leckie, Charleston, SA, 13 February 2011, personal communication).

Bob and Dawn Porter have farmed on their family property for their whole life at 'Riverside' in the Northampton Shire in WA. For many years, they carried sheep and produced wheat and lupins on their farm. When they discovered in the 1980s that their farm was not really suitable for carrying sheep due to the damage the sheep caused to the natural environment, they started to reduce sheep numbers. By 1998, all sheep had been phased out and the farm became entirely a cropping enterprise. Out of the 13,000 ha of land the farm occupies, 8000 ha are arable. Recently, the Porters took a serious look at the cropping system on their farm and decided to reduce their cropping to the most productive paddocks of around 4000 ha only. The rest has now been planted to trees (e.g. oil mallees and other native species), which are growing well.

These are just a few examples. Nationwide, most farmers are now conscious of their NRM practices. Their motivations are related closely to increasing productivity and farm sustainability. According to the survey by the ABS, nationally, 65.8% of agricultural businesses reported that they had improved their NRM practices (including the management of weeds, pests, land and soil, water and native vegetation) during 2006–2007 (ABS, 2008, p. 7). In terms of the reasons for improving NRM practices, 88.6% of agricultural businesses reported doing so to increase productivity, 88.4% for farm sustainability and 74.5% to improve environmental protection. Across all states and territories, increasing productivity and farm sustainability were the most commonly reported reasons for improving NRM practices (ABS, 2008, p. 7). Clearly, Australian farmers understand that improving their NRM practices is in their own interest, while also benefiting the whole community.

At the farm level, there are barriers to improving NRM practices. Nationally, 71.0% of agricultural businesses reported barriers to the improvement of their NRM practices during 2006–2007. Of the agricultural businesses reporting barriers, the most common reasons given were lack of financial resources (78.9%), lack of time (63.1%) and lack of government incentives (40.0%). Nationally, 22.2% of agricultural businesses gave age and/or ill health as a reason. NSW had the greatest proportion of agricultural businesses reporting barriers to improving NRM practices (74.6%), while the lowest percentage was in the NT, where 58.1% of agricultural businesses reported barriers.

9.4 Concluding Comments

People who have visited Australia are generally very impressed by its clean and beautiful environment. Australian farmers are also very proud of the products that they produce in such a 'clean and green' environment. It is fair to say that, despite the continuing existence of various environment problems, overall, Australia's environment has been well maintained and taken care of. This has benefited farmers' production and profitability enormously, and the general public as well. Australia's success in pushing for more sustainable farming can be attributed to the following major reasons.

1. **The environment-conscious public and the farming community.** The public has played a major role in pushing to protect the environment. Without their strong demand for environment protection, it is unlikely that governments would have mobilized so many resources to deal with environment-related issues. Environment education has encouraged the Australian general public to become aware of the importance of environment protection for themselves today and for their children tomorrow. In this regard, those pioneer Australian environment advocates, such as those who initiated the ACF, ought to be praised. Admittedly, environment protection cannot be carried out without a cost to farmers. This causes farmers short-term financial pain

but benefits them in the long run. In some cases, farmers are forced to take actions to care for natural resources, for example, containing chemical run-offs and preventing soil erosion. In the longer term, less polluted rivers and better-protected soils will help them to increase their profitability. In general, Australian farmers are themselves very environmentally conscious and have cooperated well with the general public to protect the natural resources on which their livelihood is based.

2. **The responsive government system.** The public's desire and effort for a better environment is not sufficient to protect the nation's environment. It is also necessary for national, state and local governments to take action to mobilize resources for this purpose and to regulate how land is used. It is not uncommon for some national governments and industry lobby groups to promote high GDP growth at the expense of the environment. Australian politicians and governments generally have been responsive to the public's demands for environment protection.
3. **The innovative approaches.** Australia has a vast land mass with diverse agricultural industries. Different regions and industries have different environment issues to care about. Australians have been very innovative in their ways of handling environment problems. The key essence is to develop different approaches to deal with environment-related problems peculiar to a region or to an industry. State and local governments have legislations or acts that focus on handling the problems under their jurisdiction, while also offering collaboration with others if the problems are interregional. Industries develop programmes to deal with environment issues related to their own industry. The innovative approach that is worth mentioning in particular is the community–government partnership, e.g. the Landcare movement. This approach has three important merits: (i) it makes good use of community resources, i.e. volunteer labour, which saves enormous labour costs that would otherwise have to be paid, while in the meantime it satisfies the desire of many individuals to make a personal contribution to protecting their environment; (ii) volunteers of various communities have great knowledge of local environment problems, which helps to develop localized approaches to deal with such problems; and (iii) because government grants are allocated to such volunteer community groups on a need basis, it ensures the resources are used to fix problems that need to be fixed or to carry out projects that will deliver greater benefits.

There remain issues related to environment protection and sustainable farming that require the attention of the general public, the farming community and the governments of Australia.

1. **Lack of the public's understanding of farmers' efforts in environment protection.** As noted earlier, Australian farmers are conscious about the importance of protecting their land and other natural resources. It is they who have been taking care of over 50% of Australia's land mass on a daily basis. However, today's urban residents do not seem to understand their problems. This lack of understanding of what is happening in the rural communities and how farmers take care of much of Australia's land mass tends to lead

them to believe that farmers are exploiters of natural resources and polluters of the natural environment. They often have misconceptions about the use of water and chemicals in some industries, such as rice, cotton and livestock. Unfortunately, urban voters collectively have a much louder voice in the public's opinions and governments often buy in to their demands, leading to legislations and acts that are impractically strict, damaging farming profitability.

2. **Lack of flexibility and practicality of some environment-related legislations and acts.** When talking to Mr Ross Leckie of SA, he made it very clear that he was generally happy with governments' responses to society's demand for environment protection. However, he pointed out that, on some occasions, some rules and regulations do make farming slightly more difficult. One example is of tree removal on the farm. On some paddocks, there may be trees that affect the efficiency of farm operations significantly. For efficiency purposes, they would like to remove some trees in the middle of a paddock and promise that for every tree they remove, they will plant significantly more, say 50, in a place that will not impede efficiency. Sometimes, the newly planted trees can also perform other functions on the farm, for example, acting as a windbreak. Unfortunately, they cannot obtain permission to do so. He believes such policies have lost their practicality (Ross Leckie, Charleston, SA, 13 February 2011, personal communication). In QLD, similar restrictions exist on the removal of trees from paddocks, causing inefficient operations. In particular, when trees close to fences are not allowed to be removed, they can cause serious damage to the fence if they are blown down by strong winds. In early 2011, the super strong cyclone Yasi caused damage to several thousands of kilometres of fence in QLD, due to falling trees. It was estimated that clearing the fallen trees alone would cost about AUS$300/km. There was also the cost to repair the fences themselves (P. Elliot, Townsville, 18 July 2011, personal communication).
3. **Lack of balance between caring for the environment and farmers' livelihoods.** Because of pressure from urban voters, some policy measures place more emphasis on protecting the environment than on farming practicality, continuity and profitability. For example, when the environment and farming compete for water, the environment may receive water with priority and that may make farming extremely difficult. One example is water sharing between the environment and farming in the Murray–Darling Basin area. In October 2010, the Murray–Darling Basin Authority released a guide to the proposed Basin Plan and held a series of information sessions with communities and the states in the Murray–Darling Basin (MDBA, 2010). Major newspapers reported that the guide to the proposed plan was met by immediate opposition from communities within the Murray–Darling Basin (Wahlquist, 2011). People in communities were concerned about the potential loss of jobs and the impacts on local communities (Morrison *et al.*, 2011). Morrison *et al.* (2011) called for a more nuanced discussion of the net benefits of sharing water in the Murray–Darling Basin.

Despite some remaining issues, Australia leads the world in many ways in regard to environmental protection and sustainable farming. It could be the case that some approaches used in Australia may not be able to be applied directly to other countries, especially some poorer developing countries. However, the philosophy of protecting the environment for sustainable farming should be the same.

Australia's success in protecting the environment and promoting sustainable farming is attributed largely to the consciousness of the society to protect the environment, governments' favourable reactions and farmers' own desire and willingness to care for the natural environment. Public involvement is the essential ingredient in the successful management of Australia's natural environment.

In view of the great value of public involvement in protecting the environment, for some high-income countries, Australia's community-based 'landcare group' model should be of great relevance. For developing countries, what they can learn from Australia's experience is slightly complicated.

Environment consciousness is somehow related to one's standard of living. If one's access to life essentials, especially food, is a problem, it is hard for one to place environment protection before one's survival. With extreme conditions in some poor developing countries, farmers do tend to exploit the land for immediate survival (to feed the family), and in such circumstances, sustainable farming may be ignored.

Caring for the natural environment costs money. Even in Australia, activities in counteracting negative impacts on the environment from farming still have to be profitable, or at least revenue neutral (Pannell, 2011). For farmers in developing countries who can barely survive, it is not realistic to expect them to spend extra money to care for the environment, unless it increases their profitability. For those financially better-off farmers, their efforts in caring for the environment still have to be financially rewarding.

Despite difficulties, caring for the environment in developing countries is still important. Governments of developing countries need to take greater responsibility for initiating and financing environmental protection programmes and for promoting sustainable farming. Public involvement through community–government partnership should also be trialled to reduce pollution of the environment in both rural and urban areas.

Notes

[1]The fire risk varies with the type of agriculture. For example, for those parts of the NT, QLD and WA in dry savannah areas under grazing, fire is a big risk if no rain follows during the dry season.

[2]Following the change in government, on 30 March 2012 the new Liberal National Party of Queensland government announced machinery-of-government changes for departments. The functions of the former Department of Environment and Resource Management will now be delivered by the following departments: Environment and Heritage Protection, Natural Resources and Mines, National Parks, Recreation, Sport and Racing, Energy and Water Supply and Science, Information Technology, Innovation and the Arts.

[3]A bioregion is defined as a large area of land characterized by its vegetation, animal communities, soil structure and climate. The division of land according to bioregion can help better resource management. All land areas in QLD are placed into various bioregions. A map showing all the bioregions in QLD can be found at: http://www.derm.qld.gov.au/vegetation/pdf/bioregion_v0001.pdf. The Australian land mass

is divided into 85 bioregions and 403 subregions. The maps of these 85 bioregions and 403 subregions can be found at http://www.environment.gov.au/parks/nrs/science/bioregion-framework/ibra/index.html.

[4]Based on the Little River Landcare Group (2011) and Red Meat Green Facts (2011). Red Meat Green Facts was created by the Australian cattle and sheep industry in an ongoing effort to discuss the issues of sustainable farming practices.

References

ABS (2000) *Year Book Australia 2000*. Australian Government, Canberra.

ABS (2008) *Natural Resource Management on Australian Farms 2006–07*. Pub No 4620.0. Australian Government, Canberra.

ANRA (2009) Rangelands overview – Australia. Australian Government (http://www.anra.gov.au/topics/rangelands/overview/index.html, accessed 26 September 2011).

Australian Government Land and Coasts (2011) Caring for our Country (http://www.nrm.gov.au/about/caring/index.html, accessed 26 August 2011).

Australian Greenhouse Office [AGO] (2011) Australian Greenhouse Office (http://www.accc.gov.au/content/index.phtml/itemId/527041/fromItemId/815972/quickLinkId/815426/whichType/org, accessed 5 September 2011).

Campbell, K.O. (1980) *Australian Agriculture: Reconciling Change and Tradition*. Longman Cheshire, Melbourne, Australia.

Caring for our Country (2011) Landcare funding from Caring for our Country (http://www.nrm.gov.au/business-plan/funded/09/landcare/landcare-nt.html, accessed 29 September 2011).

CSIRO (2007) Climate change in Australia: Technical report 2007. CSIRO, Canberra.

Department of Climate Change and Energy Efficiency (2011) Australian national greenhouse accounts: National Inventory Report 2009, Volume 1. Commonwealth of Australia, Canberra.

Department of the Prime Minister and Cabinet (2010) Interview, Prime Minister of Australia, Prime Minister of Australia's website. Department of the Prime Minister and Cabinet, Canberra, 27 April 2010 (http://pmrudd.archive.dpmc.gov.au/node/6708, accessed 5 September 2011).

DERM [Department of Environment and Resource Management] (2011) Website of Queensland Department of Environment and Resource Management (http://www.derm.qld.gov.au/, accessed 14 September 2011).

Donovan, P. (1995) *In the Interest of the Country: A History of the Pastoral Board of South Australia 1893–1993*. Pastoral Management Branch of the South Australian Department of Environment and Natural Resources, Adelaide, Australia.

Garnaut, R. (2008) *The Garnaut Climate Change Review: Final Report*. Cambridge University Press, Cambridge, UK.

Garnaut, R. (2011) *The Garnaut Review 2011: Australia in the Global Response to Climate Change*. Cambridge University Press, Cambridge, UK.

Gomez, K. (2011) drumMUSTER® and ChemClear: success through shared responsibility. Presentation at the ABARES 2011 Outlook Conference, Canberra, 1–2 March 2011 (http://www.daff.gov.au/abares/conferences-events, accessed 8 September 2011).

Landcare Australia (2011) What is Landcare? (http://www.landcareonline.com.au/about/what-is-landcare/, accessed 29 September 2011).

Little River Landcare Group (2011) Promoting a healthy, productive and diverse biological and social environment (http://www.littleriver-landcare.org.au/, accessed 27 September 2011).

MDBA (2010) *Guide to the Proposed Basin Plan*. Murray–Darling Basin Authority, Canberra.

MLA (2010) Reducing Emissions (http://www.redmeatgreenfacts.com.au/Reduce-Emission, accessed 27 September 2011).

Morrison, M.D., Wheeler, S.A. and MacDonald, D.H. (2011) Towards a more nuanced discussion of the net-benefits of sharing water in the Murray–Darling Basin. *AFBM Journal* 8, 27–38.

Pannell, D. (2011) Environment protection: challenges for future farming. *AFBM Journal* 8, 19–26.

Parliament of Australia (2003) Inquiry into Australia's Response to Global Warming: Government Members Report. Commonwealth of Australia, 28 April 2003 (http://www.aph.gov.au/senate/Committee/ecita_ctte/completed_inquiries/1999-02/gobalwarm/report/d01.htm, accessed 5 September 2011).

PBSA [Pastoral Board of South Australia] (2010) Annual Report 2009–10. Government of South Australia, Adelaide, Australia.

Preston, B. and Jones, R. (2006) Climate Change Impacts on Australia and the Benefits of Early Action to Reduce Global Greenhouse Gas Emissions: A Consultancy Report for the Australian Business Roundtable on Climate Change. CSIRO, Canberra.

Red Meat Green Facts (2011) Little River – big picture. Case studies (http://www.redmeatgreenfacts.com.au/Case-Studies/Little-River-NSW, accessed 27 September 2011).

Senate Environment, Communications, Information Technology and the Arts References Committee (2000) The heat is on: Australia's greenhouse future. Commonwealth of Australia (http://www.aph.gov.au/senate/Committee/ecita_ctte/completed_inquiries/1999-02/gobalwarm/report/contents.htm, accessed 5 September 2011).

Victorian Department of Primary Industries (2011) Landcare groups (http://www.dpi.vic.gov.au/dpi/vro/vrosite.nsf/pages/lwm_landcare_groups, accessed 29 September 2011).

Wahlquist, A. (2011) The media and the guide to the basin plan. In: Connell, D. and Grafton, R.Q. (eds) *Basin Futures: Water Reform in the Murray–Darling Basin*. Griffin Press, Canberra, pp. 115–134.

III Handling Emerging Challenges for Future Success: The Australian Way

10 Handling Emerging Challenges

As has been shown in earlier chapters, Australia has successfully developed a very modern and advanced agriculture, to which a number of factors have contributed. Other countries are envious of some dimensions of Australian agriculture. Australians feel proud of their agriculture. In this final chapter of the book, we show how Australians are handling emerging challenges for future agricultural success.

As in many other parts of the world, Australian agriculture is facing increasing challenges resulting from domestic and global economic, political, social and cultural changes. Added to these challenges are the uncertainties caused by climatic variations. The Australian way of handling some of these major challenges is highlighted in the following sections. Increasingly uncertain world markets and volatile weather conditions (especially severe floods and droughts) are two major complications that make farming in Australia most challenging. The ways in which these two challenges are handled are presented in Sections 10.1 and 10.2.

10.1 Uncertain World Markets

Over 60% of Australia's agricultural products are exported. This makes it very difficult for Australian farmers to secure a relatively stable income from selling their products, due to various obstacles: chiefly, difficulties in market access, volatile prices of agricultural products and fluctuating exchange rates.

While Australian farmers have a strong capability to produce more to supply to the world, finding suitable markets has been a huge challenge. Competition in the world agricultural market is tough. The lack of progress in multilateral trade negotiations limits market access. Australian governments at both the federal and state levels are working to help farmers gain market access in various ways, including:

- negotiating bilateral and regional free trade agreements to counteract the lack of progress of the Doha round of multilateral trade negotiations;
- promoting Australian agricultural products globally through Austrade and other trade missions;
- state governments also set trade offices in major trading countries to help farmers to sell;
- the federal government, through Austrade, provides grants to help farmers to export; and
- providing information dissemination seminars or forums to help farmers to understand overseas market developments (such as changing tastes and preferences).

Bilateral and regional free trade agreements (FTAs) have been used by the Australian Government as a major approach to help access overseas markets. Currently, Australia has six FTAs in force with New Zealand, Singapore, Thailand, the USA, Chile and (with New Zealand) the Association of South-east Asian Nations. The countries covered by these FTAs account for 28% of Australia's total trade.

The Australian Government insists that it will not enter into any trade agreement that falls short of the benchmarks set by the World Trade Organization (WTO) or the benchmarks Australia sets itself for high-quality, truly free trade deals that support global trade liberalization. Indeed, in a WTO review, the Australia–New Zealand Closer Economic Relations Trade Agreement, which took effect in January 1983, was 'recognized as the world's most comprehensive, effective and multilaterally compatible free-trade agreement'.

On 22 May 2012, the Malaysia–Australia FTA was signed. It is currently undergoing domestic approval processes and is expected to take effect on 1 January 2013. Eight more FTAs are under negotiation, including with China, Japan and the Republic of Korea. The countries covered by these eight negotiations, plus Malaysia, account for a further 44% of Australia's trade. The eight FTAs that are under negotiation are:

- Australia–China FTA.
- Australia–Gulf Cooperation Council FTA.
- Australia–India Comprehensive Economic Cooperation Agreement.
- Australia–Japan FTA.
- Australia–Korea FTA.
- Indonesia–Australia Comprehensive Economic Partnership Agreement.
- Pacific Agreement on Closer Economic Relations Plus.
- Trans-Pacific Partnership Agreement.

Currently, the Australian Government's highest regional trade negotiation priority is the conclusion of the Trans-Pacific Partnership (TPP) Agreement. This agreement builds on the Trans-Pacific Strategic Economic Partnership Agreement (P4) between Brunei Darussalam, Chile, New Zealand and Singapore, which came into force in 2006. The TPP includes the P4 parties as well as Australia,

Peru, the USA, Vietnam and Malaysia. Recently, Japan, Canada and Mexico have also formally expressed interest in joining the TPP negotiations.

Handling the volatile prices of agricultural products and foreign exchange fluctuations is another major challenge. The prices of some products can be vastly different between the time of sowing and the time of harvesting. Too low or too high a value of the Australian dollar can make exporting easier or harder. Fluctuations in exchange rates can also have detrimental effects on the actual income farmers receive in Australian dollars.

Given that, nowadays, selling products is the job of each individual farmer, some farmers have started to use various risk management tools to handle volatility in product prices and foreign exchange rates, such as forward contracts, futures, or even options, and currency swaps. Using such tools is very new and challenging to farmers. Australian governments provide training workshops and various other supporting programmes to help farmers learn such new tools; for example, Farmready under the current Australia's Farming Future programme. Various agribusiness firms which rely on farmers' financial success, such as financial institutions and supply companies, also provide similar training and advice to farmers.

10.2 Volatile Weather Conditions

There have been reports suggesting that climate change could result in more frequent occurrences of extreme weather events, such as severe floods and droughts (Preston and Jones, 2006; CSIRO, 2007; Garnaut, 2008, 2011). While floods and droughts are not new to Australian farmers, more severe weather conditions, especially severe and prolonged droughts, will be very destructive to the livelihood of many Australian farmers. According to Keogh (2012), Australian agriculture is more volatile than the agriculture of most other nations, and is the most volatile sector of the Australian economy. Volatile weather conditions, together with volatile world prices of agricultural products and exchange rates, make it likely that farm income in Australia will also become more volatile.

Farmers in Australia generally have been well prepared for adverse extreme weather conditions, due partly to their long-time experience in battling such weather extremes and partly to the proactive approaches adopted by Australian governments in preparing them to survive any harsh weather conditions. Some of the recent initiatives by Australian governments are as follows, some of which have been mentioned in earlier chapters.

- Encouraging on-farm carbon sequestration to reduce carbon emissions from farming.
- The Carbon Farming Initiative allows farmers to earn carbon credits and then to sell these credits to businesses wishing to offset their emissions.
- Allowance in the Emission Trade Scheme to help farmers trade carbon credits for income.
- Farm Management Deposit scheme to help farmers to smooth out income over good and bad years.

- The 'Pilot of Drought Reform Measures' to support farmers, their families and rural communities better in preparing for future challenges, rather than waiting until they are in crisis before offering assistance.

10.3 Biosecurity Threats

Australia enjoys a very good plant and animal health status and remains free of many major plant and animal diseases. This enviable status, however, faces a wide range of biosecurity challenges during the years to come. These challenges range from changes in disease risk, ecosystems, technology and the policy environment (Nunn, 2011).

Australia takes strict measures to protect its plant and animal health status. It has long had a strong quarantine presence at its border, with close inspection of incoming passengers, goods and mail to help prevent the entry of diseases and pests. Such strict measures have helped Australia in the past to deny the entry of many unwanted plant and animal diseases and pests. However, in recent years, diseases and pests have tended to spread much faster and travel further globally. Contributing factors include demographic changes, intensive crop and animal production, increasing globalization, growth in trade, faster urbanization and other environmental changes – including climate change – that influence the ecology of disease agents (Cohen, 2000). Hence, protecting Australia's biosecurity will become more challenging in the future.

Well coordinated measures are required to prevent and manage any outbreak of crop and animal diseases and pests. Farmers can play their part by monitoring any signs of diseases and pests closely. Recent advances in information and communications technologies have made monitoring and reporting much easier through the widespread use of 'social media' (e.g. Facebook, Twitter).

Enhanced disease surveillance systems are a key component of maintaining and improving biosecurity. In the past 10 years or so, Australia has increased its investments in this area. In addition to the strong quarantine presence at its border, the Australian Government is now also placing emphasis on pre-border biosecurity such as capacity building in animal disease control in neighbouring countries (Nunn, 2011). These activities help both to control known disease and pest risks in these countries, thus reducing the risk of spread to Australia, and also to provide early warning and intelligence of changing risks, such as from newly recognized emerging diseases. Greater emphasis is also being given to offshore audit and verification in countries that are major trading partners. These activities help manage biosecurity risks offshore and aim to ensure that what is exported to Australia is safe (Nunn, 2011).

Two other key institutions are Animal Health Australia (AHA) and Plant Health Australia (PHA). AHA is a not-for-profit public company established by the Australian Government, state and territory governments and major national livestock industry organizations. AHA manages more than 50 national programmes on behalf of its members that are designed to improve biosecurity, animal and human health, market access, livestock welfare, productivity and food safety and quality (AHA, 2012a).

PHA is also a not-for-profit organization that works in partnership with industry, governments, researchers and others to facilitate improvements in the policy, practice and performance of Australia's plant biosecurity system and to build capability to respond to plant pest emergencies. It is the lead national coordinating body for plant health in Australia. It endeavours to enhance Australia's plant health status, assist trade, safeguard the livelihood of producers and support the sustainability and profitability of Australia's plant industries and the communities that rely on them (PHA, 2012).

In early 2012, AHA and PHA signed a Memorandum of Understanding to formalize and strengthen the existing partnership between the two companies. They already had mechanisms for consultation, management and implementation of joint projects and services to members. The Memorandum of Understanding provides further support for them to strengthen the mechanisms established for their joint management of *Farm Biosecurity* programmes. It also supports their ongoing work to streamline their two emergency response deeds and to collaborate on additional biosecurity initiatives, including the newly created Biosecurity Planning Reference Group, comprising government and plant and livestock industry member representatives (AHA, 2012b).

10.4 Farm Succession

The farm succession issue is not new but its importance is increasing. Many farm children have chosen to leave the family farm, or have been encouraged to leave the farm to pursue careers elsewhere. This results in no obvious successor when family seniors decide to retire. The issue is causing concern in all parts of rural Australia, but the problem is more acute in areas where there is a long distance to travel to major centres (Stephens, 2011). Inappropriate handling of farm succession can affect on-farm productivity negatively.

Compared with other challenges, this has received relatively less attention, although key stakeholders (such as governments, business bodies and farmers themselves) are aware of the issue. Many farmers tend to leave the matter until very late, often causing hardship to those who are affected by the succession. Some farmers do not know how to address the issue and some choose not to deal with it because it is too emotional for them. Barclay *et al.* (2007) provide useful discussions about farm succession in Australia and compare the trends in Australia with some other countries.

There have been efforts to address the issue. *A Guide to Succession – Sustaining Families and Farms*, recently developed in collaboration by some rural RDCs, is one example of support available to farm businesses (Wilkinson and Sykes, 2007). This guide provides a professional perspective and ideas on the succession-planning process, case studies of family situations and checklists for identifying the important factors that need to be considered when embarking on this process.

Some large business bodies have also started to provide advice and service to their clients about how to handle the succession issue. For example, as a major institution providing financial services to Australia's rural communities, Rabobank has appointed a number of Farm Succession Managers in different parts of the country to help their clients on this important matter. The bank offers a comprehensive succession facilitation service to help farm families to deal with their succession planning (K. Lee, Farm Succession Manager, Rabobank, Orange, 21 July 2011, personal communication; Rabobank, 2012).

10.5 Labour and Skills Shortage

How to attract labour with the skills needed for farming has increasingly become a challenge. Australian agriculture faces both acute short-term and emerging chronic long-term labour and skills shortages. Acute labour and skills shortages are caused by short-term factors; for example, the strong growth in the mining sector, which has attracted labour and skills from the rural sector. The potential emergence of chronic long-lasting labour and skills shortages is related to a decline in new entrants seeking employment in the rural sector. Addressing both shortages is very important but very challenging (Nelson, 2011).

Various strategies have been proposed by Australian agricultural industries to address labour and skill shortages. In 2008, the NFF developed a Labour Shortage Action Plan in consultation with regional and sectoral industry bodies as well as government (NFF, 2008). The plan highlighted the following strategies:

- understand workforce requirements to target recruitment better;
- publicize and promote the knowledge- and technical-intensive roles that are emerging in the sector, particularly in schools, to encourage new entrants and graduates;
- develop specific strategies to overcome issues of remoteness and seasonality in different industries to ensure regional Australia can provide an environment to attract workers;
- improve career pathways and in particular encourage education and training to support skills development;
- promote the broader work and lifestyle opportunities that are associated with roles in the agricultural sector; and
- encourage flexibility and simplicity of employment arrangements and regulation to meet needs better.

Communities have also been called on to be proactive in attracting, involving and retaining skilled newcomers. In a report by Kilpatrick *et al.* (2010), they found that mobile skilled workers were drawn to, wanted to be part of and stayed in communities that were innovative, embraced diversity and accepted newcomers.

Various other initiatives have also been taken. Rural RDCs and various other rural-related organizations often provide scholarship opportunities to young people who are working or are willing to work in the future in agriculture for tertiary education. Currently, RIRDC administers an 'Investing in Youth Studentship Program', which is a collaborative initiative across industry and government. The programme offers undergraduate students financial support and a mentor who can provide career advice. It also offers the recipients relevant industry placements so they can gain experience in their chosen field of study. Recipients are selected on the basis of their commitment to agriculture (RIRDC, 2012).

10.6 Stronger Public Demand for Environment Protection

When the standard of living improves, the demand for environment protection by the broad community becomes stronger. In most countries, especially in countries where governments are elected democratically, politicians are generally very responsive to such demands, resulting in more strict rules and regulations about environment protection. This is clearly the case in Australia. Since the 1970s, when environmental concerns came to prominence (see Chapter 9), communities have become increasingly concerned about environmental issues. Politicians have responded accordingly. New policies on environment protection potentially have major implications for farming, in terms of land use, the ways in which farm operations are carried out and the extra financial burdens from operational complications.

10.6.1 Water use for environment protection

As a result of the decade-long drought in the 2000s, a strong public demand to divert water from irrigation to environmental use emerged, especially in the Murray–Darling Basin area. A draft Murray–Darling Basin Plan has been drawn up. It is proposed that water will be bought back from irrigators in the basin for environmental use. This will, however, have significant impact on irrigation farmers, either directly or indirectly (Morrison *et al.*, 2011). Farmers will need to cope with reduced water availability, thus handling the challenge of how best to adapt their management practices and land uses accordingly. They may need to keep up with water-conserving innovations, requiring time, effort and expense. In the meantime, the Australian governments are also encouraging farmers to adopt water-saving and environment-friendly practices.

10.6.2 Chemical use and environment protection

Another major public concern is about the on-farm use of chemicals. Chemical use has been a major contributor to the growth of agricultural productivity in Australia since the 1960s. Its use, however, also produces pollution of the environment, through runoffs, residuals in the soil and used containers (if not properly disposed of). Farmers in Australia generally have been very aware of the possible pollution resulting from the use of chemicals. They understand that inappropriate use of chemicals on the farm affects their sustainable use of farming resources and can even be harmful to animals and humans. None the less, they have to respond to public demand about reduced and smarter use of chemicals and the government's stricter rules and regulations about chemical use.

Farmers are doing their part to reduce or avoid pollution of the environment from chemical use. As noted earlier, farmers are adopting innovative techniques that can help in reducing runoffs and trapping sediments. There is an organized mechanism to help in recycling the containers of chemicals to reduce environment pollution. Increasingly, farmers are using precision farming techniques to reduce on-farm chemical use. In addition, the Australian Government provides funds to hold workshops to train farmers to use chemicals correctly.

10.7 Urban–Rural Divide

When the Europeans came to Australia to settle, there was no such thing as urban areas or cities; all the land was covered by native vegetation. It was only after many years of significant contributions made by many of the pioneers to develop the agricultural sector that towns and cities gradually emerged. Not surprisingly, many of the people who moved to live in the so-called urban areas had strong connections with agriculture, through personal experience, relatives or friends. In those earlier days, agriculture in Australia was viewed favourably because agriculture provided a significant share of wealth for the country. Being a farmer was a respected profession and agriculture-related university courses attracted some of the best students.

Today, however, connections between urban and rural residents have declined substantially. This is due partly to the fact that some families have lived in urban areas for a few generations and they no longer have much connection with people in rural areas and also that some new migrants with foreign urban backgrounds have no rural knowledge. Unfortunately, many such 'urban' residents have no interest in or desire to understand agriculture. According to Malcolm (2011), 'the profound ignorance of farming of well-off urban dwellers cannot be exaggerated'. Survey results support Malcolm's claim.

The Australian Council of Educational Research, commissioned by the Primary Industries Education Foundation, undertook, during June–October 2011, a national survey of students in order to understand their levels of awareness of agriculture. The survey results suggest that a high percentage of school students (around 40%) lack understanding of where food and fibre come from (Hillman and Buckley, 2011). One extreme example was that both Grade 6 and Grade 10 students had difficulty in identifying cotton socks as being a plant product, with 75% of Grade 6 students and 42% of Grade 10 students believing cotton socks were an animal product. The survey also showed that the majority of Grade 6 respondents had no family members working in primary industries.

Urban dwellers' growing ignorance of agriculture and the rural–urban disconnection has serious implications for farming and presents a new challenge to rural communities. Such disconnection leads to power imbalance between urban and rural communities, which in turn affects what farmers do and how they do it.

10.7.1 Declining rural power

The rural population in Australia has declined dramatically in the past century. This has had a significant impact on rural political power. In the past, electorates used to be weighted so as to benefit rural voters. This enabled electorates in rural areas to have fewer voters than urban electorates; an approach known as malapportionment. Today, malapportionment has been abolished in all states and rural electorates are growing larger as people move to urban areas.

Hence, representation of rural interests in parliaments has also dropped significantly. The rural–urban power imbalance causes hardship to farmers. One example is public concern for the natural environment. Ignorance of agriculture and its impact on the environment potentially could lead to legislation that could impose costs on farmers greater than the environmental gains. Farmers are less able to counter the many radical views held by urban dwellers due to the reduced relative importance of agriculture in the national economy.

10.7.2 Reconnecting rural and urban communities

Like farmers in other advanced economies where the farming population is small, Australian farmers have to live with significantly reduced rural power. Nothing they do can change the power imbalance. None the less, they can re-engage the urban population proactively to help continuously update their knowledge of agriculture. The best way to do so is through education. Some organizations, such as the Primary Industries Education Foundation and the NFF, are now engaged in educating urban residents.

Educating urban children about the importance of agriculture will be most effective. Schoolchildren are now encouraged to undertake farm visits, participate in local agricultural show competitions and visit local agricultural production facilities. Some schools have 'school garden programmes', which are used for promoting the participation of schoolchildren in some primary industry-related activities.

In addition to educating schoolchildren, educating the general urban public is also important. Some rural families are now opening, or have been prepared to open, their doors to host farm visits by urban visitors. Many Landcare groups also invite urban residents to become members. Most farmer organizations invite urban residents to join and have a member category that is designated specially for urban residents.

Most rural industry bodies, sometimes in collaboration with Australian governments, also make efforts to help connect urban and rural residents. The most recent joint venture between industries and governments to connect the urban with the rural is the 2012 Australian Year of the Farmer. The Australian Year of the Farmer celebrates the hard work of all farmers involved in producing, processing, handling and selling products across the country. It encourages more people to appreciate that Australian farmers produce fresh, safe and high-quality food and fibre to keep all Australians fed, clothed and sheltered; thus enriching the connections between urban and rural Australia.

10.8 Concluding Comments

This chapter shows that Australia is well prepared to meet the new and emerging challenges facing its agriculture. Arrangements and mechanisms are in place to mitigate possible impacts resulting from volatile world markets or severe weather conditions, or to deal with biosecurity threats. Efforts are also being made to handle several other major challenges such as farm succession, labour and skills shortage and the rural–urban divide.

Ongoing commitments are needed to deal with future challenges in agriculture. Australian governments, industry bodies and the farming community are not short of initiatives to handle future challenges.

Among all the emerging and future challenges, the one that stands out is: in the future, who is going to farm the land in Australia? In this regard, increased attention is warranted on farm succession and rural labour training. Concerted endeavours are called for to handle this challenge strategically.

To attract more talent to farming, farming as a career must be attractive. Hence, it is most imperative for Australia to maintain a high rate of productivity growth through investments in R&D, education and public infrastructure so that Australian agriculture maintains its comparative advantage in the domestic economy and relative to its international competitors. This in turn will also enable Australia to continue its role as a major food producer for the world, given that there will be three billion more people to feed in the next few decades.

To close this book, Australian agriculture is well prepared for future challenges. Its preparedness for these future challenges is most amazing and impressive. With such preparedness, plus the many institutional arrangements that are conducive to agricultural development, there is every reason to believe that Australian agriculture is well positioned for future success.

References

Animal Health Australia [AHA] (2012a) About us (http://www.animalhealthaustralia.com.au/about-us/, accessed 13 April 2012).

AHA (2012b) AHA and PHA Memorandum of Understanding (http://www.animalhealthaustralia.com.au/news/2012/aha-and-pha-memorandum-of-understanding/, accessed 13 April 2012).

Barclay, E., Foskey, R. and Reeve, R. (2007) *Farm Succession and Inheritance – Comparing Australian and International Trends*. A report for the Rural Industries Research and Development Corporation, RIRDC Publication No 07/066, Canberra.

Cohen, M.L. (2000) Changing patterns of infectious disease. *Nature* 406, 762–767.

CSIRO (2007) Climate Change in Australia: Technical Report 2007. CSIRO, Canberra.

Garnaut, R. (2008) *The Garnaut Climate Change Review: Final Report*. Cambridge University Press, Cambridge, UK.

Garnaut, R. (2011) *The Garnaut Review 2011: Australia in the Global Response to Climate Change*. Cambridge University Press, Cambridge, UK.

Hillman, K. and Buckley, S. (2011) Food, Fibre and the Future: Report on Surveys of Students' and Teachers' Knowledge and Understanding of Primary Industries. The Australian Council of Educational Research (http://research.acer.edu.au/national_surveys/1, accessed 9 April 2012).

Keogh, M. (2012) Including risk in enterprise decisions in Australia's riskiest businesses. Paper presented at the 56th Annual Conference of the Australian Agricultural and Resource Economics Society, Perth, Australia, 8–10 February 2012.

Kilpatrick, S.K., Vitartas, P., Homisan, M. and Johns, S. (2010) The Mobile Skilled Workforce: Optimising Benefits for Rural Communities. Rural Industries Research and Development Corporation Publication No 10/077, Canberra.

Malcolm, B. (2011) Changing business environment: implications for farming. *AFBM Journal* 8, 73–78.

Morrison, M.D., Wheeler, S.A. and Hatton MacDonald, D. (2011) Towards a more nuanced discussion of the net-benefits of sharing water in the Murray–Darling Basin. *AFBM Journal* 8, 27–38.

National Farmers' Federation [NFF] (2008) Labour shortage action plan, Canberra, Australia (http://www.nff.org.au/get/702.pdf, accessed July 2011).

Nelson, S. (2011) Finding tomorrow's agricultural workforce. *AFBM Journal* 8, 47–60.

Nunn, M. (2011) Australia's biosecurity: future challenges for animal industries. *AFBM Journal* 8, 39–46.

Plant Health Australia [PHA] (2012) Our company (http://www.planthealthaustralia.com.au/go/phau/our-company/documents, accessed 13 April 2012).

Preston, B. and Jones, R. (2006) Climate Change Impacts on Australia and the Benefits of Early Action to Reduce Global Greenhouse Gas Emissions: A Consultancy Report for the Australian Business Roundtable on Climate Change. CSIRO, Canberra.

Rabobank (2012) Succession planning (http://www.rabobank.com.au/Rural/Succession-facilitation/Pages/Succession-planning.aspx, accessed 25 May 2012).

RIRDC (2012) Investing in Youth Studentship Program (http://www.rirdc.gov.au/programs/national-rural-issues/dynamic-rural-communities/investing-in-youth-studentship-program, accessed 18 March 2012).

Stephens, M. (2011) Planning farm succession: how to be successful. *AFBM Journal* 8, 61–66.

Wilkinson, J. and Sykes, L. (2007) A guide to succession – sustaining families and farms. GRDC report, Canberra.

Postscript

Writing this book has not been as easy as imagined initially. It required a fair amount of time, effort and persistence. However, I enjoyed the whole process and I learned a lot. Importantly, Australia's experience in developing its agriculture helped me to: (i) shift my paradigms in various ways; (ii) ask questions I otherwise would not have asked; and (iii) refine my thoughts on the essence of what is needed to develop a country's agriculture. Recorded below are a few thoughts that I am eager to share with others.

1. Government intervention in agricultural production and marketing is actually very harmful to farmers and to a country's economic prosperity

I was brought up in an agrarian society with the economy heavily controlled by the government. I was educated and I accepted that agriculture was a special industry and hence special treatments were needed. Over many years, this influenced my way of thinking about agriculture. Now, I realize that agriculture is not and should not be treated as a special industry, in both developed and developing countries. Hence, government intervention is not warranted to the extent still observed in many countries. Australia's experience provides a great example of the benefits to a developed country from not viewing agriculture as a special industry but as a component of the whole economy.

When there is significant intervention in agriculture in developing countries, three situations may apply: (i) milking agriculture; (ii) controlling prices of agricultural products, chiefly foods; and (iii) genuinely helping farmers to gain higher income. In all three cases, farmers suffer and the whole economy suffers: (i) when agriculture is milked, needless to say, farmers are disadvantaged; (ii) often, governments of developing countries intervene in agriculture in the name of stabilizing food prices, with a hidden or clear objective to please urban dwellers (in such a situation, even if farmers are provided with assistance, such as subsidized fertilizer, chemicals, farm machines and irrigation water, they still lose because such assistance may be far less than the higher output prices prevailing in world markets); and (iii) even if a government is genuine in helping farmers receive higher incomes by setting floor prices higher than the market equilibrium, farmers will still lose out in the long run. This is simply because such a situation cannot last long. Australia's reserve price scheme for wool is an example. Hence, intervention to help farmers in many developing countries is an illusion. As a result of regulations and controls, farmers are kept poor; in the long run, a country's economic prosperity is hindered.

2. Farmers are poor because urban people want them poor

In many developing countries, farmers are poor. Many reasons are responsible for this, including government's discriminatory policies against farmers. However, one important reason is that urban people want cheap food. In some countries where urban dwellers have enjoyed higher incomes, they still demand their governments keep food prices artificially low. These urban dwellers do not understand that by keeping farmers poor, their own long-run economic well-being is actually hurt, despite some short-term gains. When farmers are poor, their ability to buy goods produced in the urban system is dampened. This in turn affects the employment prospects and incomes of urban people, and ultimately hinders a country's economic development. Perhaps also, they do not appreciate that subsidizing food prices has opportunity costs in terms of other services governments could provide.

3. The right institutions and adequate levels of agricultural R&D investment are the two most important ingredients to develop a country's agriculture

There is no shortage of written work about the various strategies needed to develop a country's agriculture. The study of the experiences of Australia's agricultural development has led me to believe that if a country can get its institutions right and have an adequate level of agricultural R&D investment, its agriculture will develop well. Farmers everywhere are smart, they know what they are doing and they do not need governments to tell them what to produce and how to sell. However, they need suitable institutions to protect their rights to farm and the appropriate technologies available for them to improve their farm operations.

If a government really wants to develop its agriculture, what it needs to do is to make sure the country's institutions are conducive for farmers to farm and to give priority to investment in agricultural R&D. Institutional guarantees are needed to ensure agricultural R&D investment is continued, but not at the expense of other budget items. After all, an agrarian society needs to produce food and fibre for its citizens and hence justification for sustained agricultural R&D investment to improve a country's food and fibre production is fairly obvious.

4. If institutions are conducive, farmers around the globe can produce enough food for the world

There were no farming activities in Australia 220 years ago, but today, Australia produces a large amount of food and fibre for Australians, and also for overseas consumers, thanks to its institutional arrangements conducive to agricultural development and its very innovative farming community. This suggests that elsewhere potential exists to increase agricultural output. If such potential was brought to reality, we should have sufficient food for everyone on the planet by 2050, even if the world population increased by another three billion.

Indeed, in many parts of the world today, that people do not have enough food to eat is not because farmers in those countries cannot produce it, but is because of a lack of institutions that encourage farmers to produce. Rulers of some countries, who live luxuriously and never have food security concerns personally, could not care less about the sufferings of ordinary people. To keep power, they would rather engage in internal conflicts or civil wars. Consequently, the fields are full of mines and bullets come from nowhere. In such hostile environments, how can farmers farm the land? Where can their country people get food to eat?

Hence, if we really want to make a contribution to develop our agriculture in a country or globally, we need to urge national governments first to reform by establishing and supporting institutions that ensure the political, social and economic stability of the country. Farmers should not be discriminated against and they should be entitled to adequate representation in a country's social and political systems.

Without farmers, we have no food to eat. So, respect farmers; support farmers!

Zhang-Yue Zhou
30 June 2012
Townsville, Australia

Index

Note: Italic page numbers indicate figures and tables; numbers in brackets preceded by *n* refer to endnotes.